Lindsay

Committee on Amino Acids
Food and Nutrition Board
National Research Council

IMPROVEMENT OF PROTEIN NUTRITURE

NATIONAL ACADEMY OF SCIENCES

WASHINGTON, D.C. 1974

This study was supported in part by the
National Institutes of Health, United States Public Health Service,
Grant No. 5 RO 1 AM-05296.

Library of Congress Cataloging in Publication Data

National Research Council. Committee on Amino Acids.
    Improvement of protein nutriture.

    Includes bibliographies.
    1. Protein deficiency. 2. Protein metabolism. 3. Proteins in human nutrition. 4. Amino acid metabolism. I. Title. [DNLM: 1. Dietary proteins. 2. Amino acids. QU55 N277i]
    RC627.P7N37   1974   616.3'99   74-18462
    ISBN 0-309-02234-7

Available from
Printing and Publishing Office, National Academy of Sciences
2101 Constitution Avenue, Washington, D.C. 20418

Printed in the United States of America

Preface

The Committee on Amino Acids of the Food and Nutrition Board was initially established to consider the merit of amino acid fortification of foodstuffs in the American diet. This study led to a bulletin entitled "Evaluation of Protein Nutrition," in which information about amino acid requirements, amino acid intakes of the U.S. population, and effects of amino acid deficiencies was summarized and the nutritional merit of amino acid fortification, primarily lysine fortification, of cereal products was assessed.

The mission of the committee was subsequently broadened to include assessment of amino acid fortification as a procedure for improving protein nutriture worldwide. Almost a decade has been devoted to this study, which has been augmented by the fact that, beginning in 1971, three members of the committee have served on the FAO / WHO Expert Committee on Energy and Protein Requirements. In this way, documents generated by the FAO/WHO group became available before the final version of this report was prepared and have been particularly useful in preparing the section on nitrogen and amino acid requirements. Understandably, the two reports show distinct similarities, as well as certain differences. The FAO/WHO report develops recommendations for practical protein allowances. By contrast, this document evaluates procedures for improving protein nutriture through fortification of diets with amino acids but does not contain recommendations for protein allowances. Assessment of the validity of information on protein and amino acid requirements is obviously a necessary part of such an evaluation. Finally, although all members of the committee participated in the discussion of each section, it is presented as a collection of individual papers.

ALFRED E. HARPER, *Chairman*
D. MARK HEGSTED, *Member*
Committee on Amino Acids

FOOD AND NUTRITION BOARD

Lloyd J. Filer, Jr., *Chairman*

H. N. Munro, *Vice-Chairman*

Richard H. Barnes

Myrtle L. Brown

David L. Call

Doris H. Calloway

David Baird Coursin

E. M. Foster

Lavell M. Henderson

Robert M. Kark

Gilbert A. Leveille

Jean Mayer

Robert O. Nesheim

Nevin S. Scrimshaw

Theodore B. Van Itallie

COMMITTEE ON AMINO ACIDS

Alfred E. Harper, *Chairman*

George G. Graham

D. Mark Hegsted

L. Emmett Holt, Jr.

Esther F. Phipard

Harold H. Williams

Contents

A. E. HARPER

Basic Concepts

We tend to oversimplify the complex problem of protein nutrition in man—to seek a single figure for protein requirement and a definite value for its nutritional quality, to seek a simple explanation for the problem of "protein" malnutrition and a simple solution for improving protein nutriture. However, because nutritional needs change with age and are influenced by the environment; because mammals have complex regulatory systems that respond to the stimuli that impinge upon them; and, above all, because the body is a highly adaptable mechanism, simple answers and simple solutions to these problems are, almost without exception, inadequate.

Using the experimental approach, in which each of the many variables that influence some property of a system can be studied in turn while the others are held constant, information can gradually be assembled. In practice, the relative importance of the various factors that can be examined experimentally in isolation is difficult to assess. Such assessment must ordinarily be made through inference from information gained experimentally. This consideration of protein nutriture improvement will begin with a general discussion of basic concepts and the characteristics of regulatory systems of the body; then those topics that bear most directly on proposals for the improvement of protein nutrition—protein and amino acid requirements, methods of protein evaluation, effects of

1

deficiencies and disproportions of amino acids and the distribution of amino acids in diets—will be presented in more detail. The concluding section will deal with the practical aspects of improving protein nutriture.

THE ESSENTIALITY OF AMINO ACIDS

Man, as well as other monogastric mammals, must have enough protein to provide nitrogen and certain specific amino acids for the synthesis of new tissue during growth and reproduction, of milk constituents during lactation, and of many nitrogen-containing compounds, and for replacement of the nitrogen that is continuously lost from the body. The ability of a food protein to satisfy these requirements is determined primarily by its amino acid composition (Hegsted, 1964a, b; Meister, 1965). Because man is unable to synthesize from other compounds nine of the amino acids needed for tissue synthesis, the latter must be provided, preformed, in the diet. Amino acids that can be synthesized in the body are termed "dispensable" or "nonessential"; those that cannot are termed "indispensable" or "essential." Classification of amino acids in this way applies only to dietary need—all are essential for the synthesis of proteins.*

The amino acids that are essential for man are listed in Table 1 (Rose, 1957). Histidine, which is essential for the infant (Snyderman *et al.*, 1963), is classified in the same manner, even though adult man has not exhibited the need for a dietary supply of it (Rose, 1957). Various authorities have questioned the dispensability of histidine for adult maintenance on both theoretical and experimental grounds (Steele, 1952; Nasset and Gatewood, 1954; Nasset, 1956; Anonymous, 1964). Although microorganisms can synthesize histidine from adenylic acid, such a process is not known to occur in man (Meister, 1965). In view of the potential reserves of histidine in hemoglobin and in muscle carnosine, it is understandable that histidine might not be required for maintenance of nitrogen equilibrium in short-term studies; Rose and Wixom (1955c), however, found no adverse effects from its absence during a period of 60 days. Nonetheless, until definitive proof to the contrary is obtained, histidine should be considered an essential amino acid.

*The term "dispensable" is used here for amino acids that can be synthesized by the body. This eliminates the need for terms such as "nonessential" and "semiessential," which are inappropriate and inaccurate. It also avoids confusion from use of the term E/N, which can be taken to mean either essential/nonessential or essential/total nitrogen, by substitution of E/D for the first of these uses.

**TABLE 1   Amino Acids That Are
Essential for Man**

Isoleucine	Phenylalanine
Leucine	Threonine
Lysine	Tryptophan
Methionine	Valine
Histidine	

Some amino acids can be synthesized in the animal body, but not by all species at a rate compatible with normal growth. These may be dispensable for one species but essential for another. Arginine is apparently dispensable for man, even for the growing infant (Holt, 1967), but is essential for the young rat (Rose *et al.*, 1948). Glycine is essential for the chick but not for mammals (Meister, 1965).

A few of the dispensable amino acids can be synthesized only from specific essential amino acids. If the former are provided in the diet, the need for the amino acids from which they can be synthesized is reduced. Cystine can be formed only from methionine; and when cystine is present in the diet in adequate amounts, less methionine is required (Rose and Wixom, 1955a). Also, tyrosine can be formed only from phenylalanine; and when tyrosine is present in the diet, less phenylalanine is required (Rose and Wixom, 1955b).

The other dispensable amino acids can be synthesized in the body from organic acids that are intermediates in carbohydrate metabolism, e.g., α-ketoglutarate and pyruvate (Steele, 1952), nitrogen from surpluses of individual amino acids, and even from such compounds as ammonium citrate (Rogers *et al.*, 1970). For maximum growth of the young rat, most individual sources of nonspecific nitrogen are inferior to a mixture of all of the dispensable amino acids; and diets containing only the indispensable amino acids, even in quantities well above the requirements, do not support rapid growth (Stucki and Harper, 1962). Whether this is true for the human infant has not been established. Nitrogen for adult man can be provided in large part by glycine and diammonium citrate (Terroine *et al.*, 1930; Rose and Wixom, 1955c); but nitrogen balance in man is adversely affected when only one or two sources of nonspecific nitrogen make up a large part of the nitrogen in an amino acid diet, possibly because the rate of synthesis of the dispensable amino acids is not adequate (Swendseid *et al.*, 1960; Anderson *et al.*, 1969).

Although authorities do not completely agree on this point, the weanling rat apparently requires asparagine or glutamine for maximum growth (Breuer *et al.*, 1964; Ranhotra and Johnson, 1965; Rogers and Harper, 1965). The superiority of intact proteins over protein hydrolysates as a nitrogen source for this species during the phase of rapid growth appears to be due, at least in part, to the absence of amides from the hydrolysates. In addition, evidence indicates that, if the infant rat is fed a diet containing amino acids instead of protein from about 18 days of age, it requires another factor, as yet unidentified (Schwartz, 1970), that is associated with proteins but is not a component of them.

Besides being necessary for the synthesis of the proteins that must be formed during growth, maintenance, reproduction, and lactation, amino acids are needed for the formation of many nitrogen-containing, nonprotein substances (e.g., creatine, choline, glutathione, heme, epinephrine, and thyroxine, to name but a few) that are required for body functions. Amino acids consumed in excess of these needs undergo transamination or are deaminated, primarily but not exclusively, in the liver (Miller, 1962) to yield other amino acids or α-keto acids and ammonia. The ammonia is converted to urea, which is excreted in the urine. The α-keto acids may be used in the synthesis of fat, glucose, and other substances or may be oxidized directly to carbon dioxide and water to yield utilizable energy.

DIGESTION OF PROTEINS

The role of the gastrointestinal tract in the metabolism of proteins has been extensively reviewed (Gitler, 1964; Munro, 1964; Rogers and Harper, 1966; and Fauconneau and Michel, 1970) and documented. The proteins of foods cannot be absorbed as such, but must first undergo hydrolysis. The hydrolysis of proteins is initiated in the stomach under the influence of the gastric protease, pepsin. This enzyme is an endopeptidase, i.e., it facilitates the splitting of peptide bonds within the protein molecule. It acts most rapidly on peptide bonds in which the carbonyl group of the bond is from an aromatic amino acid. It also acts on other bonds, but less rapidly. Thus, gastric digestion of proteins results mainly in the formation of shorter polypeptides rather than in the release of free amino acids.

In addition to functioning as the initial site of protein digestion, the stomach acts as an important regulator of metabolism by controlling the rate of flow of the partially digested food into the intestine, where digestion is completed and the products of digestion are absorbed. This regulation normally prevents overloading of the intestine and limits the rate at which products of digestion pass to sites of metabolic activity within the body. The rate of stomach emptying is influenced by many factors, one of which is the composition of the diet. A high fat intake is well known to delay stomach emptying; a high protein intake also delays stomach emptying. The relative solubilities of the different components of the diet may result in differential emptying rates of different components of the diet; for example, the casein of milk precipitates out in the dilute acid solution of the gastric secretions and may be held longer in the stomach than are other, more soluble, substances. The gradual emptying of protein from the stomach, especially after ingestion of a high-protein meal, is probably important for efficient utilization of amino acids within the body.

The partially digested food, mixed with saliva and gastric juice (chyme), is emptied gradually from the stomach into the small intestine, where it is further mixed with bile, pancreatic and intestinal secretions, and cells sloughed from the intestinal mucosa. The polypeptides undergo further hydrolysis catalyzed by proteases and peptidases from the pancreas (trypsin, chymotrypsin, and carboxypeptidases) and the small intestine (probably a mixture of peptidases). The intestinal peptidases, like disaccharidases, appear to act primarily within the mucosal cells on short peptides that pass through the cell membrane (Fern *et al.*, 1969). Some proteins, such as gelatin, have peptide bonds that are very resistant to proteolytic digestion; and short peptides containing such bonds may be absorbed intact and excreted, as such, in the urine (Smiley and Ziff, 1964). These observations indicate that the high efficiency of the digestive process and the fact that many proteins are completely digested to free amino acids are accounted for, in turn, by the fact that proteins are released from the stomach at a rate compatible with complete digestion rather than that peptides are prevented from being absorbed.

Animal studies show that intestinal digestion of protein entering from the stomach is rapid. Carbon from ingested ^{14}C-labeled proteins may be found in expired carbon dioxide within 5 to 10 min after ingestion of a meal containing an isotopically labeled protein, before most of the chyme has emptied from the stomach (Hansson, 1959; Tarver, 1963). In man nitrogen from isotopically labeled yeast protein administered orally appeared in urine as rapidly as that from a yeast protein hydrolysate, and labeled nitrogen level was high in blood within 30 min (Crane and Neuberger, 1960). Most of the proteins that are secreted into the gastrointestinal tract are also digested, but usually more slowly than food proteins (Snook, 1965); and the amino acids from them are reabsorbed. The indigestible food residue, together with the undigested portion of the endogenous secretions and bacterial proteins, pass to the colon and are excreted in the feces.

DIGESTIBILITY OF FOOD PROTEINS

A measure of the apparent digestibility of a food protein can be obtained by subtracting the amount of nitrogen excreted in the feces from the amount ingested and expressing this value as a percentage of the intake:

$$\text{Apparent digestibility} = \frac{I - F}{I} \times 100,$$

where I = nitrogen intake and F = fecal nitrogen. As fecal nitrogen includes nitrogen from bacterial and sloughed intestinal mucosal cells, as well as that of undigested food proteins, apparent digestibility is not an accurate measure of the digestibility of the dietary protein. Fecal nitrogen of nondietary origin may be equivalent in amount to a large fraction of the ingested nitrogen when protein intake is low, but to only a small fraction when protein intake is high. Thus, values for apparent digestibility of a protein will vary with protein intake.

To determine true digestibility, it is necessary to correct for the amount of fecal nitrogen excreted when the subject is consuming either a protein-free diet or just enough of a highly digestible protein to prevent excessive loss of body protein. These two procedures give comparable values for fecal nitrogen of metabolic origin (Mitchell and Carman, 1926). True digestibility can then be calculated as follows:

$$\text{True digestibility} = I - \frac{(F - F_m)}{I} \times 100,$$

where F_m = fecal metabolic nitrogen loss determined as described above. The amount of fecal metabolic nitrogen is not invariable but is influenced by the amount of bulk in the diet and by the total amount of food consumed (Mitchell, 1924, 1942; Nasset, 1968). It is therefore important, in determining true digestibility, to take these factors into consideration (Hegsted, 1964a).

Most food proteins, with some exceptions, are highly digestible, 90 percent or greater. Severely heat-processed proteins, for example, may be altered in such a way that their digestibility is reduced (Liener, 1958). Some amino acids, e.g., lysine, may be altered by heat, particularly in the presence of carbohydrate, so that they do not become available to the body during digestion. Also, some peptide bonds are resistant to hydrolysis, particularly bonds involving proline. They may be hydrolyzed slowly and result in delayed release of certain amino acids. The significance of a delay in release is not known, but it may result in some reduction of the efficiency of utilization of the absorbed amino acids. When animals are fed a diet deficient in a single amino acid, which is provided separately a few hours later, protein utilization is not improved nearly as much as it is by supplementation of the diet itself; the response is greater with some amino acids than with others (Elman, 1939; Geiger, 1948; Yang *et al.*, 1968). This situation, however, is more extreme than that likely to be encountered as a result of slow digestibility of proteins in ordinary diets.

Finally, considerable controversy has occurred about the nutritional significance of the protein contributed by digestive secretions and sloughed mucosal cells (Gitler, 1964; Munro, 1964; Nasset, 1968; Fauconneau and Michel, 1970). They may contribute to the intestinal contents during a 24-h period at least as much, if not substantially more, protein as is normally consumed by the average American during the same time span. The proteins that are secreted into the gastrointestinal tract, however, must be resynthesized by the body from amino acids that are absorbed from the intestine. As the amount of nitrogen excreted in feces is normally small and very constant, the reutilization of proteins secreted into the gastrointestinal tract represents, primarily, a recycling of amino acids of metabolic or endogenous origin. Hence, even though the process may contribute substantially to the rapid turnover of amino acids in the body and represent an important component of the dynamic state of body proteins, its nutritional contribution is taken into account by the methods used for estimating amino acid and protein requirements.

ABSORPTION OF AMINO ACIDS

Free amino acids are the end products of protein digestion. A few small peptides may not be hydrolyzed; as mentioned above, some proline-containing peptides from gelatin are absorbed and excreted in the urine (Smiley and Ziff, 1964). Otherwise, amino acids enter the blood stream in free form.

Extensive study of intestinal absorption of amino acids (Wilson, 1962; Christensen, 1963; Stein, 1967) has well established that the naturally occurring L-amino acids are transported across the intestinal wall by an active process. This process, which transports amino acids against a concentration gradient, requires energy; it is inhibited by metabolic inhibitors, such as cyanide and dinitrophenol, and by a lack of oxygen. It has kinetic properties comparable to those of enzyme systems, indicating that it is a carrier-mediated process. Different amino acids are absorbed at different rates—the dicarboxylic amino acids are transported most slowly. When the concentrations of amino acids are high, some will compete with others for transport systems; and mutual inhibitory effects may be observed. On the basis of work with isolated ascites tumor cells, evidence from measurements of competitive inhibition suggests that several distinct systems for the transport of amino acids exist and that each is more or less specific for a certain group (Oxender and Christensen, 1963). Most of the systems studied are sodium-dependent (Christensen *et al.*, 1967). Transport systems across membranes are probably similar throughout the body (Stein, 1967; Christensen, 1968).

The question arises whether amino acid absorption ever becomes a limiting factor in the nutrition of the normal subject. Regulation of stomach emptying and the continuous nature of protein digestion ensure that the intestine will not ordinarily be overloaded and that free amino acids will be released gradually. Also, the intestine is a large organ with an immense surface area, so its capacity for absorption is great. Therefore, even if competition among amino acids for absorptive sites occurs, this should be short-lived and unlikely to reduce the efficiency of amino acid utilization. Studies with rats showed no evidence that a large excess of leucine in the diet impaired the absorption of isoleucine (Rogers and Harper, 1968). Also, observations on the absorption of mixtures of amino acids from tied-off segments of rat intestines containing about the amounts that would be expected after a meal did not reveal evidence of competitive inhibition during absorption (Gitler, 1964). The overall efficiency of digestion and absorption is exemplified by observations that the apparent digestibility of wheat gluten by rats did not decrease when the amount in the diet was increased to as much as 75 percent (Munaver and Harper, 1959) and that the amount of leucine excreted in rat feces remains small, even when the amount in the diet is great enough to be detrimental (Tannous, 1963).

The gastrointestinal tract is thus a highly efficient system for releasing and extracting the amino acids of ingested food proteins. It also contributes to conservation of amino acids by controlling their rate of flow into the blood stream through regulation of gastric emptying.

DISTRIBUTION AND METABOLISM OF AMINO ACIDS

Absorbed amino acids pass from intestine to the portal vein and directly to the liver. Some amino acids may be removed for the regeneration of intestinal proteins during absorption. Intestinal tissues contain transaminases that are particularly active with glutamic acid as a substrate. Thus, the quantity of this amino acid may decrease and that of alanine increase during absorption (Neame and Wiseman, 1958; Peraino and Harper, 1963; Pion *et al.*, 1964). Despite the large quantity of protein secreted into the intestinal tract and presumably digested and reabsorbed (Snook, 1965; Fauconneau and Michel, 1970), an abnormal dietary pattern of amino acids in a meal is reflected in the free amino acid pattern of the portal blood (McLaughlan and Morrison, 1968). The abnormal free amino acid pattern may be evident throughout the entire blood supply for several hours after a meal, if the meal is large (McLaughlan *et al.*, 1963); but in the post-absorptive state, the plasma amino acid pattern returns to the rather stable fasting pattern (Adibi, 1968; Coulson and Hernandez, 1968; McLaughlan and Morrison, 1968). If the meal is small, or the protein content of the diet low, the blood amino acid pattern may not reflect that of the diet, as amino acids will be rapidly removed from blood by tissues for protein synthesis (Anderson and Linkswiler, 1969).

The concentrations of amino acids in tissues ordinarily exceed those in blood (Tallan *et al.*, 1954; Herbert *et al.*, 1966), indicating that amino acids are actively transported across cell membranes. The rate of transport increases as the concentration of the amino acid in the medium increases (Stein, 1967); therefore, when the concentrations of amino acids in blood rise after a meal, the rate of entry into tissues should also increase. Amino acid uptake by muscle is stimulated by insulin (Wool *et al.*, 1968) and into liver by glucagon (Mallette *et al.*, 1969; Tews *et al.*, 1970). Because amino acids stimulate secretion of both insulin and glucagon (Fajans *et al.*, 1967; Ohneda *et al.*, 1968), an influx of amino acids into the bloodstream would stimulate amino acid uptake as well.

The liver is the main organ of protein catabolism and an important site for the synthesis of glycine, serine, alanine, and aspartic and glutamic acids. It withdraws amino acids from the blood passing through it for the synthesis of both liver and plasma proteins; contributes amino acids that have been synthesized; oxidizes them, especially those in surplus; and incorporates the nitrogen into urea (Tarver, 1963). Little quantitative information is available about the impact on absorbed amino acids of passage through the liver (Elwyn, 1970); but studies of systemic blood amino acids in man and experimental animals indicate that, for some time after ingestion of a meal containing an unbalanced pattern of amino acids, this pattern is reflected in the systemic blood (McLaughlan and Morrison, 1968). There are probably two reasons for this: First, the composite amino acid pattern of liver and plasma proteins resembles that of a well-balanced dietary protein, so that removal of amino acids for protein synthesis in the liver will tend to reduce the concentration of an amino acid that is in short supply; second, surpluses of ingested amino acids that cannot be used for protein synthesis will pass through the liver, if the capacity of the liver for amino acid oxidation is exceeded (Yoshida *et al.*, 1966; Benevenga *et al.*, 1968). Many of the enzymes for the catabolism of indispensable amino acids in liver are not highly active unless protein intake has been high (Harper, 1965), so the removal of amino acids that are in surplus requires time. Elwyn *et al.* (1968) concluded from observations in dogs fed a single meal that the liver may take up the entire output of amino acids from the intestine. Presumably this would occur only if the supply of protein or energy, or both, were low.

Amino acids not removed by the liver enter the general circulation and are transported to various other organs and tissues. In these, too, the pattern of amino acids required for protein synthesis resembles that of a well-balanced dietary protein; any abnormalities in pattern will remain evident in the circulating blood until regulatory mechanisms come into play to correct them. Again, this is particularly true if one amino acid is in short supply, because a greater proportion of it will be removed from the body fluids and its concentration will fall. This is observed regularly after the feeding of diets with imbalances of amino acids or deficient in a single one of them (McLaughlan and Illman, 1967). Substantial quantities of some amino acids will accumulate in muscle cells (Young, 1970); and transamination, particularly of dispensable amino acids, can lead to alanine formation in muscle (Pozefsky *et al.*, 1969). However, since most of the indispensable amino acids are oxidized in the liver, surpluses of these are removed only after recirculation to this organ. The branched-chain amino acids are unique among the indispensable amino acids in undergoing transamination in muscle (Harper *et al.*, 1970).

Different organs and tissues incorporate amino acids from the circulating body fluids into proteins at different rates. The pancreas and small intestine, and probably most secretory organs, incorporate them rapidly, as shown by experiments with isotopically labeled amino acids (Tarver, 1963). Ingested amino acids taken up by the pancreas, for example, may be detected in the proteins of pancreatic secretions within 1–2 h (Hansson, 1959). The liver also takes up amino acids rapidly, but less so than some of the primarily secretory organs. Muscle and brain incorporate amino acids more slowly, owing to their slower average rate of protein turnover. Nevertheless, the total quantity of amino acids taken up by muscle may be great because it represents such a large proportion of the total body mass.

Amino acids circulating through the kidney are efficiently reabsorbed from the glomerular filtrate, as evidenced by the low loss of them in the urine of animals fed a high-protein diet (Sauberlich and Baumann, 1946). Even when the circulating concentration of leucine is in the range that leads to adverse effects, urinary loss is small (Tannous, 1963). The kidney serves as an organ of excretion of amino acids only when the renal threshold is exceeded; and this appears to occur with tyrosine, for example, only when the tyrosine load is great enough to cause debilitation of the animal (Boctor, 1967). As with the intestine, the kidney appears to have evolved as an organ for conservation of amino acids. It does not function as an effective regulatory mechanism for removal of excessive amounts of amino acids, even when circulating concentrations are in the toxic range.

Munro (1970) has extensively discussed amino acid pools. The amounts of free amino acids in organ and tissue pools are small in relation to the requirements of the growing organism; furthermore, the pattern of amino acids in these pools bears little relationship to the pattern of amino acid requirements. The free amino acids of plasma and liver are in a highly dynamic state (Henriques *et al.*, 1955; Black, 1968; Haider and Tarver, 1969) and turn over rapidly; those of muscle equilibrate less readily with other pools (Henriques *et al.*, 1955). Muscle, by virtue of its mass, is the largest reservoir of free amino acids; and, as the latter tend to accumulate when plasma concentrations rise after a meal, the muscle pool may serve as a temporary storehouse from which they can be released later, when plasma concentrations have fallen. This could reduce the flow of amino acids to sites of oxidation in the liver after a large meal has been consumed.

The body does not store extra amino acids as it does energy sources. There is no compartment in the body with a reserve of protein that is comparable to the reserves of carbohydrate, as glycogen, in liver and muscle, or of fat in adipose tissue (Holt and Halec, 1962). The proportion of protein in the body (as a percentage of lean body mass) remains very constant after the third year of life in individuals consuming an adequate diet (Munro, 1964). Although protein is lost from the body during periods of starvation or protein deficiency, surfeit consumption of protein, even a very large surplus, results in little accumulation in the body of the rat (Mayer and Vitale, 1957). Lack of amino acid storage is also emphasized by observations that the growth of young organisms ceases within hours (Bender, 1965) after they begin consuming a diet that is deficient in a single essential amino acid. In adult man, negative nitrogen balance, indicating loss of body protein, occurs within 24 h. Finally, if a diet lacks only one essential amino acid, which is provided several hours later, efficient use of all amino acids falls (Elman, 1939; Geiger, 1948; Spolter and Harper, 1961).

Although no actual storage of amino acids occurs, the ability of the body to maintain some organs at the expense of others during depletion has given rise to the concept of "protein reserves." This term, however, is controversial (Holt and Halec, 1962); some investigators prefer the term "labile body proteins," which, although more appropriate, still leaves something to be desired. Regardless of the term, most authorities agree about the nature of the phenomenon. During the initial phase of starvation or consumption of a low-protein or protein-free diet, rapid loss of body protein occurs, particularly from the liver and alimentary tract. Thereafter the loss is slow; but eventually, upon prolonged starvation or protein depletion, muscle is severely depleted and contributes much more to the total protein loss than do the visceral organs. The implication of these observations is that amino acids released from the breakdown of proteins that are not crucial for survival are reutilized to maintain the essential structures and enzymes of the body (Waterlow, 1968). Brain proteins and oxidative enzymes are only slowly depleted.

What may be overlooked, however, is that some proteins are labile in animals fed a protein-free diet but not in animals subjected to starvation, even though both conditions lead to protein depletion. Many of the enzymes of amino acid catabolism in liver fall to very low levels in animals fed a protein-free diet but increase in activity during starvation. Hence, it appears that the lability of body proteins is a reflection of adaptive processes, which will be discussed later, and that whether a specific protein is labile or nonlabile may depend upon the specific nutritional or physiologic state of the organism (Harper, 1965; Waterlow, 1968). How much protein can be lost during periods of depletion, without impairing the ability of the body to respond or adapt to a new challenge, will depend upon the severity of the conditions; but adaptation to under nutrition must at some point shade into deterioration.

NITROGEN BALANCE AND THE DYNAMIC STATE OF PROTEIN METABOLISM

Nitrogenous compounds, primarily proteins, are ingested continuously throughout the life of an individual; and nitrogen is continuously ex-

creteu in the urine and feces and is lost to a lesser extent from the skin.
Even when the diet contains no protein, the loss of body nitrogen is
continuous—after an initial rapid loss of nitrogen for a few days, the
urinary or endogenous loss tends to stabilize at about 2 mg of nitrogen/
basal kcal/day for several species of animals. It is somewhat lower than
this for man (see H. H. Williams *et al.*, pp. 23–63 in this volume).
Fecal and cutaneous losses occur in addition to this. An amount of
protein sufficient to replace that lost must therefore be consumed daily
to maintain an individual in nitrogen equilibrium. Balance studies (Irwin
and Hegsted, 1971) have provided evidence that the minimal nitrogen
requirement exceeds the estimates of nitrogen losses.

Measurement of deviations from nitrogen equilibrium (i.e., nitrogen
balance) by determining the intake and output of nitrogen can give an
estimate of the overall metabolic state of a subject (Allison, 1964;
Hegsted, 1964a, b). Nitrogen balance (B) is calculated as follows:

$$B = I - (U + F + S),$$

where I is nitrogen intake, U is nitrogen excreted in the urine, F is
nitrogen excreted in the feces, and S is nitrogen lost from the skin or
integuments. S is commonly disregarded in short-term nitrogen balance
measurements, but it should be noted that this can cause a substantial
error in longer studies (Holmes, 1965; Sirbu *et al.*, 1967). When B is
positive, nitrogen intake exceeds nitrogen excretion; hence, the body is
gaining nitrogen. The body should be in positive balance during growth
or during repletion of depleted tissues. When B is negative, nitrogen intake is less than nitrogen excretion, and the body is losing nitrogen.
Negative balance is associated with depletion of body proteins. When
intake just equals excretion, B is zero and the body is said to be in nitrogen equilibrium. Balance, however, represents the sum of gains and
losses from all tissues of the body; it does not give any idea of the relative gains or losses of protein by different tissues or organs.

For many years it was believed that nitrogen metabolism could be
divided into endogenous and exogenous metabolism (Folin, 1905).
Endogenous metabolism was considered to be highly constant and independent of diet and was equated with the nitrogen lost in urine,
mainly in creatinine and uric acid, by an individual consuming a protein-
free diet. Exogenous metabolism was considered to be associated with
the utilization of dietary protein and was equated with nitrogen excreted in the urine, primarily as urea, which fluctuated in response to
protein intake. This gave rise to the idea that body proteins were highly
stable and that, during maintenance, protein was required primarily for

the replacement of "worn-out" cells, for the formation of proteins that were secreted or excreted, and for the formation of nonprotein nitrogenous substances that were lost in the urine. Any excess of dietary proteins was assumed to be directly catabolized without entering into these processes (Folin, 1905).

Studies with isotopically labeled amino acids have shown that this concept is inaccurate and that most tissue proteins undergo continuous degradation and synthesis, even in subjects who are in nitrogen equilibrium (Schoenheimer, 1942; Neuberger and Richards, 1964). Protein metabolism is thus a dynamic process in which amino acids from the diet are mixed in the body fluids with amino acids that arise from the hydrolysis of tissue proteins to form a common pool—or more probably pools—from which new tissue proteins are resynthesized. This process of degradation and resynthesis continues even when no protein is being ingested. A very useful process it is; for, during periods of dietary deprivation, it enables the body to utilize amino acids from proteins that are not essential for survival, especially muscle proteins, for the maintenance of such essential organs as the heart, liver, and kidneys. If the amino acids released during the breakdown of tissue proteins were not efficiently recaptured, the protein requirements of mammals would be immense (Fauconneau and Michel, 1970).

The major fate of amino acids from diets containing moderate amounts of well-balanced proteins is incorporation into tissue proteins. Obviously, the system for incorporating absorbed amino acids into tissue proteins is highly efficient. Otherwise, diets containing limited amounts of high-quality protein would not be utilized so efficiently for growth. One factor probably contributing to this is that many amino acid-degrading enzymes are present in small amounts in the liver unless protein intake exceeds the requirement. In addition, the rates of reaction of these enzymes are low if their substrate concentrations are low. The rates of reaction of protein-synthesizing systems are high even when amino acid concentrations are low, and the effective amino acid concentration for the maximum rate of these reactions is much below that required for amino acid-degrading enzymes.

Protein-synthesizing systems are also responsive to the supply of amino acids. The capacity of cell-free systems from the liver to synthesize protein increases as a direct function of the amino acid supply of the liver from which they are isolated (Jefferson and Korner, 1969). Munro (1968, 1970) and Sidransky *et al.* (1970) have shown that the integrity of the protein-synthesizing system depends upon the presence of an adequate supply of all amino acids and that if tryptophan, in particular, is lacking, the system fails to aggregate. In rats fed a protein-deficient diet, the rate of protein synthesis in muscle falls rapidly

(Waterlow, 1968). With more prolonged deficiency, the activities of the amino acid-activating enzymes in the liver increase (Gaetani *et al.*, 1964), as though this organ were primed for trapping available amino acids. Thus amino acid supply evidently affects regulation of the utilization of amino acids for protein synthesis at the subcellular level.

The various body proteins do not all turn over at the same rate (Neuberger and Richards, 1964). Most collagen proteins, for example, turn over very slowly—some scarcely at all. At the other extreme are the proteins of the digestive secretions, which must be synthesized, degraded, and resynthesized within a matter of hours. Since the entire intestinal mucosa is replaced within 2-4 days, protein turnover in this tissue is obviously rapid. The turnover rates of various other proteins fall between these extremes and are not always constant, but may be influenced by the endocrine and nutritional state of the organism (Waterlow, 1968).

HOMEOSTASIS AND ADAPTATION

Living organisms are dynamic systems with complex regulatory mechanisms that function to maintain a relatively constant internal environment (the blood and body fluids) despite changes in the external environment that tend to alter it. This ability, homeostasis, is a basic characteristic of living systems. Food is a component of the external environment that, upon digestion and absorption, impinges upon and may alter the internal environment, thus bringing homeostatic mechanisms into play. If the food consumed provides the various nutrients in approximately the amounts and proportions required by the organism, the internal environment is restored to its standard state within a short time after ingestion of a meal. If, on the other hand, the composition or the quantity of food consumed deviates appreciably from that required, the capacity of the regulatory mechanisms may be exceeded. Unless the organism undergoes adaptations that improve its ability or increase its capacity to restore the internal environment to the standard state, deleterious effects may ensue.

The basic homeostatic mechanisms regulate the transport of nutrients and their metabolic products and the activities of enzyme systems involved in the utilization of nutrients and their metabolic products. In higher animals, the coordination of these processes throughout the body is accomplished through the endocrine and nervous systems. In the organism existing under relatively stable conditions, e.g., consuming a relatively constant diet for a prolonged period of time, the regulatory mechanisms tend to stabilize at a level of activity that can readily counteract moderate alterations in the internal environment. As an illustration of this, the rate of reaction of an enzyme is influenced by the concentration of substrate, end products, activators, and inhibitors, so that, without any change in the concentration of the enzyme, the rate of the reaction it catalyzes responds to moderate changes in the flow of substrate. If, however, the diet is drastically altered, the capacity of the enzyme system may be exceeded and the internal environment may be altered for a prolonged period of time. Unless the enzyme system undergoes an adaptative increase in capacity, the organism may be affected adversely.

The capacity to adapt is another type of homeostatic mechanism. In response to a drastic change in the internal environment, the concentration of an enzyme system may increase owing to an enhanced rate of enzyme synthesis or a decreased rate of degradation, or it may decrease owing to a decreased rate of enzyme synthesis or an increased rate of degradation (Schimke and Doyle, 1970).

These homeostatic mechanisms—both constitutive regulation and adaptation—are important in relation to modifications in the composition or content of the dietary protein and in relation to evaluation of protein nutrition. The organism can tolerate a wide range of protein intakes. As intake increases above the amount required to maintain the various body structures in their standard state, the excess of amino acids is degraded and used as a source of energy. If intake increases enough, the capacity of the organism for degradation of amino acids may be exceeded; and amino acids will accumulate in body fluids. As a result, entry of amino acids into the body may be slowed by a reduction in the rate of stomach emptying; and, if the protein content of the diet is high enough, entry may be further decreased by a reduction in voluntary food intake. If protein intake remains elevated and the concentrations of amino acids in body fluids remain high, the liver and kidneys enlarge; the concentrations of many of the enzymes of amino acid degradation increase, i.e., the animal undergoes adaptations that enable it to adjust to the high protein intake; and homeostasis is restored at an elevated level of protein metabolism.

If protein intake is low, as a result of either starvation or consumption of a protein-free or low-protein diet, other adaptations occur. With low protein but adequate caloric intake, the activities of many amino acid-degrading enzymes fall to low levels—a state compatible with the conservation of the essential amino acids but not with the disposal of a large excess of them should the protein intake suddenly be greatly increased (Harper, 1965). Also, turnover of body proteins becomes slower, another mechanism that tends to conserve amino acids (Waterlow, 1968).

Under conditions of starvation, on the other hand, even though protein is lacking, the most immediate need is for glucose, which cannot be synthesized from the fatty acids released through mobilization of adipose tissue fat but only from amino acids (Cahill *et al.*, 1966). In this state, the amino acid-degrading enzymes do not fall, as they do in the animal fed a low-protein diet; and some are known to increase in activity. The enzymes involved in gluconeogenesis also tend to increase. Hence, adaptation in this case is directed toward providing the precursors of glucose for the nervous system rather than toward conserving amino acids. Adjustments or adaptations of differing magnitude occur with different feeding regimens (Knox and Greengard, 1965; Kaplan and Pitot, 1970). They occur in response to alterations in endocrine status (Kenney, 1970) and during infections and stresses. Despite our limited knowledge of adaptive phenomena, it is clear that overall protein utilization is altered by them.

Thus, protein nutrition cannot be viewed accurately as a set of recommendations for some standard or average state. The nutritional value of a protein is not a constant. It depends upon the physiological state and the protein and energy status of the individual. Plasma amino acids tend to be very constant in the fasting state but to fluctuate in relation to food consumption and the composition of the dietary protein. Requirements change gradually with age, nutritional status, and health factors. The effects of disproportionate amounts of amino acids vary with diet and age. All of these factors must be considered in relation to problems of improving protein nutriture and of obtaining the most efficient utilization of protein resources. It is not possible to provide figures that take into account all of these eventualities. The available figures and recommendations should be viewed as guides from which extrapolations must be made, using common sense and keeping firmly in mind the dynamic state of the organism and its capacity to undergo adaptations that favor survival.

REFERENCES

Adibi, S. A. 1968. Influence of dietary deprivations on plasma concentration of free amino acids of man. J. Appl. Physiol. 25:52–57.

Allison, J. B. 1964. The nutritive value of dietary proteins. Pages 41–84 *in* H. N. Munro and J. B. Allison, eds. Mammalian protein metabolism. Vol. 2. Academic Press, New York.

Anderson, H. L., and H. Linkswiler. 1969. Effect of source of dietary nitrogen on plasma concentrations and urinary excretion of amino acids of men. J. Nutr. 99:91–100.

Anderson, H. L., M. B. Heindel, and H. Linkswiler. 1969. Effect on nitrogen balance
    of adult man of varying source of nitrogen and level of calorie intake. J. Nutr.
    99:82-90.
Anonymous. 1964. Histidine requirement in infancy. Nutr. Rev. 22:114-115.
Bender, A. E. 1965. The balancing of amino acid mixtures and proteins. Proc. Nutr.
    Soc. 24:190-196.
Benevenga, N. J., A. E. Harper, and Q. R. Rogers. 1968. Effect of an amino acid
    imbalance on the metabolism of the most limiting amino acid in the rat. J. Nutr.
    95:434-444.
Black, A. L. 1968. Modern techniques for studying the metabolism and utilization
    of nitrogenous compounds, especially amino acids. Pages 287-309 in Isotope
    studies on the nitrogen chain. International Atomic Energy Agency, Vienna.
Boctor, A. M. 1967. Some nutritional and biochemical aspects of tyrosine toxicity
    and lysine availability. Ph.D. Thesis. Massachusetts Institute of Technology,
    Cambridge.
Breuer, L. H., Jr., W. G. Pond, R. G. Warner, and J. K. Loosli. 1964. The role of
    dispensable amino acids in the nutrition of the rat. J. Nutr. 82:499-506.
Cahill, G. F., Jr., M. G. Herrera, A. P. Morgan, J. S. Soeldner, J. Steinke, P. L. Levy,
    G. A. Reichard, Jr., and D. M. Kipnis. 1966. Hormone-fuel interrelationships
    during fasting. J. Clin. Invest. 45:1751-1769.
Christensen, H. N. 1963. Amino acid transport and nutrition. Fed. Proc. 22:1110-
    1114.
Christensen, H. N. 1968. Relevance of transport across the plasma membrane to the
    interpretation of the plasma amino acid pattern. Pages 40-52 in J. H. Leathem,
    ed. Protein nutrition and free amino acid patterns. Rutgers University Press, New
    Brunswick, N.J.
Christensen, H. N., M. Liang, and E. G. Archer. 1967. A distinct Na^+-requiring
    transport system for alanine, serine, cysteine, and similar amino acids. J. Biol.
    Chem. 242:5237-5246.
Coulson, R. A., and T. Hernandez. 1968. Amino acid catabolism in the intact rat.
    Am. J. Physiol. 215:741-746.
Crane, C. W., and A. Neuberger. 1960. The digestion and absorption of protein by
    normal man. Biochem. J. 74:313-323.
Elman, R. 1939. Time factor in retention of nitrogen after intravenous injection of
    a mixture of amino-acids. Proc. Soc. Exp. Biol. Med. 40:484-487.
Elwyn, D. 1970. The role of liver in regulation of amino acid and protein metabo-
    lism. Pages 523-556 in H. N. Munro, ed. Mammalian protein metabolism. Vol. 4.
    Academic Press, New York.
Elwyn, D., H. Hamendra, C. Parikh, and W. C. Shoemaker. 1968. Amino acid move-
    ments between gut, liver and periphery in unanesthetized dogs. Am. J. Physiol.
    215:1260-1275.
Fajans, S. S., J. C. Floyd, Jr., R. F. Knopf, and J. W. Conn. 1967. Effect of amino
    acids and proteins on insulin secretion in man. Pages 617-662 in G. Pincus, ed.
    Recent progress in hormone research. Vol. 23. Academic Press, New York.
Fauconneau, G., and M. C. Michel. 1970. The role of the gastrointestinal tract in
    the regulation of protein metabolism. Pages 481-522 in H. N. Munro, ed. Mam-
    malian protein metabolism. Vol. 4. Academic Press, New York.
Fern, E. B., R. C. Hider, and D. R. London. 1969. The sites of hydrolysis of dipep-
    tides containing leucine and glycine by rat jejunum in vitro. Biochem. J. 114:
    855-861.

Folin, O. 1905. A theory of protein metabolism. Am. J. Physiol. 13:117–138.

Gaetani, S., A. M. Paolucci, M. A. Spadoni, and G. Tommassi. 1964. Activity of amino acid activating enzymes in tissues from protein-depleted rats. J. Nutr. 84: 173–178.

Geiger, E. 1948. The role of the time factor in feeding supplementary proteins. J. Nutr. 36:813–819.

Gitler, C. 1964. Protein digestion and absorption in nonruminants. Pages 35–69 in H. N. Munro and J. B. Allison, eds. Mammalian protein metabolism. Vol. 1. Academic Press, New York.

Haider, M., and H. Tarver. 1969. Effect of diet on protein synthesis and nucleic acid levels in rat liver. J. Nutr. 99:433–445.

Hansson, E. 1959. The formation of pancreatic juice proteins studied with labelled amino acids. Acta Physiol. Scand. 46, Suppl. 161.

Harper, A. E. 1965. Effect of variations in protein intake on enzymes of amino acid metabolism. Can. J. Biochem. 43:1589–1603.

Harper, A. E., N. J. Benevenga, and R. M. Wohlheuter. 1970. Effects of ingestion of disproportionate amounts of amino acids. Physiol. Rev. 50:428–558.

Hegsted, D. M. 1964a. Proteins. Pages 115–179 in G. H. Beaton and E. W. McHenry, eds. Nutrition. Vol. 1. Academic Press, New York.

Hegsted, D. M. 1964b. Protein requirements. Pages 135–172 in H. N. Munro and J. B. Allison, eds. Mammalian protein metabolism. Vol. 2. Academic Press, New York.

Henriques, O. B., S. B. Henriques, and A. Neuberger. 1955. Quantitative aspects of glycine metabolism in the rabbit. Biochem. J. 60:409–424.

Herbert, J. D., R. A. Coulson, and T. Hernandez. 1966. Free amino acids in the caiman and rat. Comp. Biochem. Physiol. 17:583–598.

Holmes, E. G. 1965. An appraisal of the evidence upon which recently recommended protein allowances have been based. World Rev. Nutr. Diet. 5:237–274.

Holt, L. E., Jr. 1967. Amino acid requirements of infants. Curr. Ther. Res. 9(Suppl.): 149–156.

Holt, L. E., Jr., and E. Halac, Jr. 1962. The concept of protein stores and its implications in diet. J. Am. Med. Assoc. 181:699–705.

Irwin, I., and D. M. Hegsted. 1971. A conspectus of research on protein requirements of man. J. Nutr. 101:385–430.

Jefferson, L. S., and A. Korner. 1969. Influence of amino acid supply on ribosomes and protein synthesis of perfused rat liver. Biochem. J. 111:703–712.

Kaplan, J. H., and H. C. Pitot. 1970. The regulation of intermediary amino acid metabolism in animal tissues. Pages 388–436 in H. N. Munro, ed. Mammalian protein metabolism. Vol. 4. Academic Press, New York.

Kenney, F. T. 1970. Hormonal regulation of synthesis of liver enzymes. Pages 131–172 in H. N. Munro, ed. Mammalian protein metabolism. Vol. 4. Academic Press, New York.

Knox, W. E., and O. Greengard. 1965. The regulation of some enzymes of nitrogen metabolism—an introduction to enzyme physiology. Adv. Enzyme Regul. 3:247–313.

Liener, I. E. 1958. Effect of heat on plant proteins. Pages 79–129 in A. M. Altschul, ed. Processed plant protein foodstuffs. Academic Press, New York.

McLaughlan, J. M., and W. I. Illman. 1967. Use of free plasma amino acid levels for estimating amino acid requirements of the growing rat. J. Nutr. 93:21–24.

McLaughlan, J. M., and A. B. Morrison. 1968. Dietary factors affecting plasma amino acid concentrations. Pages 3–18 *in* J. H. Leathem, ed. Protein nutrition and free amino acid patterns. Rutgers University Press, New Brunswick, N.J.

McLaughlan, J. M., F. J. Noel, A. B. Morrison, and J. A. Campbell. 1963. Blood amino acid studies. IV. Some factors affecting plasma amino acid levels in human subjects. Can. J. Biochem. Physiol. 41:191–200.

Mallette, L. E., J. H. Exton, and C. R. Park, 1969. Effects of glucagon on amino acid transport and utilization in the perfused rat liver. J. Biol. Chem. 244:5724–5728.

Mayer, J., and J. J. Vitale. 1957. Thermochemical efficiency of growth in rats. Am. J. Physiol. 189:39–42.

Meister, A. 1965. The role of amino acids in nutrition. Pages 201–230 *in* Biochemistry of the amino acids. Vol. I. Academic Press, New York.

Miller, L. L. 1962. The role of the liver and non-hepatic tissues in the regulation of free amino acid levels in the blood. Pages 708–721 *in* J. T. Holden, ed. Amino acid pools. Elsevier, New York.

Mitchell, H. H. 1924. A method for determining the biological value of protein. J. Biol. Chem. 58:873–903.

Mitchell, H. H. 1942. An evaluation of feeds on the basis of digestible and metabolizable nutrients. J. Anim. Sci. 1:159–173.

Mitchell, H. H., and G. G. Carman. 1926. The biological value of the nitrogen of mixtures of patent white flour and animal foods. J. Biol. Chem. 68:183–215.

Munaver, S. M., and A. E. Harper. 1959. Amino acid balance and imbalance. II. Dietary level of protein and lysine requirement. J. Nutr. 69:58–64.

Munro, H. N. 1964. The role of the gastrointestinal tract in protein metabolism. Blackwell, Oxford. 402 pp.

Munro, H. N. 1968. Role of amino acid supply in regulating ribosome function. Fed. Proc. 27:1231–1237.

Munro, H. N. 1970. Free amino acid pools and their role in regulation. Pages 299–386 *in* H. N. Munro, ed. Mammalian protein metabolism. Vol. 4. Academic Press, New York.

Nasset, E. S. 1956. Essential amino acids and nitrogen metabolism. Pages 3–21 *in* W. H. Cole, ed. Some aspects of amino acid supplementation. Rutgers University Press, New Brunswick, N.J.

Nasset, E. S. 1968. Contribution of the digestive system to the amino acid pool. Pages 80–87 *in* J. H. Leathem, ed. Protein nutrition and free amino acid patterns. Rutgers University Press, New Brunswick, N.J.

Nasset, E. S., and V. H. Gatewood. 1954. Nitrogen balance and hemoglobin of adult rats fed amino acid diets low in L- and D-histidine. J. Nutr. 53:163–176.

Neame, K. D., and G. Wiseman. 1958. The alanine and oxo acid concentrations in mesenteric blood during the absorption of L-glutamic acid by the small intestine of the dog, cat, and rabbit *in vivo*. J. Physiol. 140:148–155.

Neuberger, A., and F. F. Richards. 1964. Protein biosynthesis in mammalian tissues. Pages 243–296 *in* H. N. Munro and J. B. Allison, eds. Mammalian protein metabolism. Vol. 1. Academic Press, New York.

Ohneda, A., E. Parada, A. M. Eisentraut, and R. H. Unger. 1968. Characterization of response of circulating glucagon to intraduodenal and intravenous administration to amino acids. J. Clin. Invest. 47:2305–2322.

Oxender, D. L., and H. N. Christensen. 1963. Distinct mediating systems for the transport of neutral amino acids by the Ehrlich cell. J. Biol. Chem. 238:3686–3699.

Peraino, C., and A. E. Harper. 1963. Observations on protein digestion *in vivo*. V. Free amino acids in blood plasma of rats force-fed zein, casein or their respective hydrolysates. J. Nutr. 80:270–278.

Pion, R., G. Fauconneau, and A. Rerat. 1964. Variation de la composition en acides amines du sang porte au cours de la digestion chez le porc. Ann. Biol. Anim. Biochim. Biophys. 4:383–401

Pozefsky, T., P. Felig, J. O. Tobin, J. S. Soeldner, and G. F. Cahill, Jr. 1969. Amino acid balance across tissues of forearm in postabsorptive man. Effects of insulin at two dose levels. J. Clin. Invest. 48:2273–2282.

Ranhotra, G. S., and B. C. Johnson. 1965. Effect of feeding different amino acid diets on growth rate and nitrogen retention of weanling rats. Proc. Soc. Exp. Biol. Med. 118:1197–1201.

Rogers, Q. R., and A. E. Harper. 1965. Amino acid diets and maximal growth in the rat. J. Nutr. 87:267–273.

Rogers, Q. R., and A. E. Harper. 1966. Protein digestion: Nutritional and metabolic considerations. World Rev. Nutr. Diet. 6:250–291.

Rogers, Q. R., and A. E. Harper. 1968. Significance of tissue pools in the interpretation of changes in plasma amino acid concentrations. Pages 107–126 in J. H. Leathem, ed. Protein nutrition and amino acid patterns. Rutgers University Press, New Brunswick, N.J.

Rogers, Q. R., D. M-Y. Chen, and A. E. Harper. 1970. Importance of dispensable amino acids for maximum growth in the rat. Proc. Soc. Exp. Biol. Med. 13:517–522.

Rose, W. C. 1957. The amino acid requirements of adult man. Nutr. Abstr. Rev. 27:631–647.

Rose, W. C., and R. L. Wixom. 1955a. The amino acid requirements of man. XIII. The sparing effect of cystine on the methionine requirement. J. Biol. Chem. 216:763–773.

Rose, W. C., and R. L. Wixom. 1955b. The amino acid requirements of man. XIV. The sparing effect of tyrosine on the phenylalanine requirement. J. Biol. Chem. 217:95–101.

Rose, W. C., and R. L. Wixom. 1955c. The amino acid requirements of man. XVI. The role of the nitrogen intake. J. Biol. Chem. 217:997–1004.

Rose, W. C., M. J. Oesterling, and M. Womack. 1948. Comparative growth on diets containing ten and nineteen amino acids with further observations upon the role of glutamic and aspartic acid. J. Biol. Chem. 176:753–762.

Sauberlich, H. E., and C. A. Baumann. 1946. The effect of dietary protein upon amino acid excretion by rats and mice. J. Biol. Chem. 166:417–428.

Schimke, R. T., and D. Doyle. 1970. Control of enzyme levels in animal tissues. Annu. Rev. Biochem. 39:929–976.

Schoenheimer, R. 1942. The dynamic state of body constituents. Harvard University Press, Cambridge, Mass. 78 pp.

Schwarz, K. 1970. An agent promoting growth of rats fed amino acid diets (Factor G). J. Nutr. 100:1487–1500.

Sidransky, H., D. S. R. Sarma, M. Bongiorno, and E. Verney. 1970. Effect of dietary tryptophan on hepatic polyribosomes and protein synthesis in fasted mice. J. Biol. Chem. 243:1123–1132.

Sirbu, E. R., S. Margen, and D. H. Calloway. 1967. Effect of reduced protein intake on nitrogen loss from the human integument. Am. J. Clin. Nutr. 20:1158–1165.

Smiley, J. D., and M. Ziff. 1964. Urinary hydroxyproline excretion and growth. Physiol. Rev. 44:30–44.

Snook, J. T. 1965. Dietary regulation of pancreatic enzyme synthesis, secretion and inactivation in the rat. J. Nutr. 87:297–305.

Snyderman, S. E., A. Boyer, E. Roitman, and L. E. Holt, Jr. 1963. The histidine requirement of the infant. Pediatrics 31:786–801.

Spolter, P. D., and A. E. Harper. 1961. Utilization of injected and orally administered amino acids by the rat. Proc. Soc. Exp. Biol. Med. 106:184–189.

Steele, R. 1952. The formation of amino acids from carbohydrate carbon in the mouse. J. Biol. Chem. 198:237–244.

Stein, W. D. 1967. The movement of molecules across cell membranes. Academic Press, New York. 369 pp.

Stucki, W. P., and A. E. Harper. 1962. Effects of altering the ratio of indispensable to dispensable amino acids in diets for rats. J. Nutr. 78:278–286.

Swendseid, M. E., C. L. Harris, and S. G. Tuttle. 1960. The effects of sources of non-essential nitrogen on nitrogen balance in young adults. J. Nutr. 71:105–108.

Tallan, H. H., S. Moore, and W. H. Stein. 1954. Studies on the free amino acids and related compounds in the tissues of the cat. J. Biol. Chem. 211:927–939.

Tannous, R. I. 1963. Metabolic studies on leucine, isoleucine and valine antagonism in the rat. D.Sc. Thesis. Massachusetts Institute of Technology, Cambridge.

Tarver, H. 1963. Metabolism of amino acids and proteins. Pages 449–548 in Ch. Rouiller, ed. The liver. Academic Press, New York.

Terroine, E. F., R. Bonnet, R. Chotin, and G. Mourot. 1930. Le rôle des sels ammoniaçaux organiques, des albumines déficientes et des albumines efficaces dans la converture de la dépense azotée endogène spécifique. Arch. Int. Physiol. 33:60–85.

Tews, J. K., N. A. Woodcock, and A. E. Harper. 1970. Stimulation of amino acid transport in rat liver slices by epinephrine, glucagon and adenosine 3′, 5-monophosphate. J. Biol. Chem. 245:3026–3032.

Waterlow, J. C. 1968. Observations on the mechanism of adaptation to low protein intakes. Lancet 2:1091–1097.

Wilson, T. H. 1962. Intestinal absorption. W. B. Saunders, Philadelphia. 263 pp.

Wool, I. G., W. S. Stirewalt, K. Kurihara, R. B. Low, P. Bailey, and D. Oyer. 1968. Mode of action of insulin in the regulation of protein biosynthesis in muscle. Pages 139–213 in E. B. Astwood, ed. Recent progress in hormone research. Vol. 24. Academic Press, New York.

Yange, S. P., K. S. Tilton, and L. L. Ryland. 1968. Utilization of a delayed lysine or tryptophan supplement for protein repletion of rats. J. Nutr. 94:178–184.

Yoshida, A., P. M-B. Leung, Q. R. Rogers, and A. E. Harper. 1966. Effect of amino acid imbalance on the fate of the limiting amino acid. J. Nutr. 89:80–90.

Young, V. R. 1970. The role of skeletal and cardiac muscle in the regulation of protein metabolism. Pages 585–674 in H. N. Munro, ed. Mammalian protein metabolism. Vol. 4. Academic Press, New York.

H. H. WILLIAMS, A. E. HARPER, D. M. HEGSTED,
G. ARROYAVE, and L. E. HOLT, JR.

Nitrogen and Amino Acid Requirements

The dietary need for nitrogen or protein includes a highly specific requirement for essential amino acids and a nonspecific requirement that can be met by a variety of nitrogenous compounds (Terroine, in Waterlow and Stephen, 1957). The quality of a food protein depends upon the amounts and proportions of the essential amino acids that it provides; the quantity of food needed to meet protein requirements depends upon its total nitrogen content and upon its content of each essential amino acid.

Animal cells do not store amino acids as they do carbohydrates and fats. Instead, they use only as a source of energy those consumed in excess of immediate needs only. Nitrogen, also, is lost continuously in sloughed skin, hair, and nails and in urine, feces, sweat, and various other body secretions and excretions. The supply of nitrogen and amino acids for tissue synthesis must, therefore, be replenished continuously from the diet.

Since the latter part of the 19th century, the nitrogen requirement of adult man has been repeatedly estimated, initially through epidemiologic observations and later through nitrogen balance experiments and measurements of nitrogen losses. During the past 25 yr, amino acid and nitrogen requirements of infants, children, men, and women have been quantified. Nitrogen requirements for growth and maintenance under standard conditions are well established. Nevertheless, much remains to be learned about factors that influence these requirements. Amino acid requirements, even for growth and maintenance, are less well established than those for nitrogen.

ESTIMATION OF NITROGEN REQUIREMENTS

Two ways of estimating nitrogen or food protein requirements are the so-called "factorial method" and the nitrogen balance method.

The factorial method involves summation of four components: (1) endogenous losses of nitrogen in urine; (2) metabolic losses in feces; (3) losses through dermal and other minor routes; and (4) deposits in new tissue during growth or reproduction or that secreted in milk during lactation. The dermal losses include not only the nitrogen in sweat, but also that in desquamated epithelium, hair, and nails (Mitchell and Edman, 1962). Small amounts are also lost from menstrual, seminal, nasal, and oral secretions and excretions from wounds. The fourth component, "true growth," is important in infants and children until maturity and during pregnancy. An additional demand for dietary nitrogen occurs during lactation for the formation of milk proteins, and during convalescence from illness or protein deprivation for the repletion of tissue proteins.

Holt *et al.* (1962) have concluded that there is no virtue in protein intakes in excess of the quantity required to provide for these needs. However, estimation of nitrogen requirements from summation of the amounts of nitrogen lost and amounts of protein synthesized is based on the assumption that the nitrogen consumed can be used with 100 percent efficiency. Egg and other high-quality proteins are used with efficiencies approaching this value when they are fed in limited amounts to growing animals, but evidence from several studies indicates that even the highest-quality proteins are not used as efficiently by man when fed at about the requirement level.

If precise values were available for each of the various categories of nitrogen losses and for efficiency of nitrogen utilization in replacing them, an accurate estimate of dietary nitrogen or food protein requirements could be made. An approximation of the minimum nitrogen requirement can be calculated from available figures for losses and accretion, based on what appear to be reasonable assumptions, but information on efficiency of nitrogen utilization can be obtained only through direct measurements in nitrogen balance studies. In view of the variability among individuals (coefficient of variation for these measurements is commonly 15–20 percent), of the changes in requirements with age and physiological state, and of the many factors that can influence nitrogen retention, it is doubtful whether more precise estimates of nitrogen losses than those now available would make estimates of requirements more meaningful.

The second method of estimating nitrogen requirements involves measurements of the least amount of nitrogen from the highest-quality proteins that will maintain nitrogen equilibrium in adults or satisfactory growth and nitrogen retention in children. The nitrogen balance method is used for this; and, although it too has several shortcomings (Smuts, 1935; Nasset, 1956; Wallace, 1959; Hegsted, 1964), it does give a direct estimate of nitrogen needs and of efficiency of utilization of different proteins in maintaining nitrogen equilibrium.

Unless efficiency of nitrogen utilization is 100 percent, values obtained by the nitrogen balance method should exceed those obtained by the factorial method. A correction for any inefficiency of nitrogen utilization must be included when estimates of nitrogen requirements are made by the factorial method.

Requirements may be reported in various ways—as the maximum for any individual, as the mean or median, or as the mean adjusted for variability within a population. Each of these has its particular merit. It is important to recognize that the nitrogen and amino acid requirements of individuals, like other nutritional requirements, may differ substantially and that average values, assuming a normal distribution, exceed the requirements for one-half of the population and fail to meet those of the other half. Recommended dietary allowances are calculated to exceed the requirements of almost all individuals and thus are considerably in excess of the average requirement. The subsequent discussion is concerned with average requirements, not with allowances. The values presented should, therefore, not be confused with recommended dietary allowances.

THE FACTORIAL APPROACH AND ENDOGENOUS NITROGEN LOSSES

Endogenous Urinary Nitrogen Excretion

Smuts (1935), on the basis of studies of several animal species, proposed a value of 2 mg/basal kcal as the minimum amount of nitrogen excreted in urine (endogenous nitrogen) by subjects who had been ingesting a protein-free diet for several days. This value has been used in estimating the protein requirements of man (Hegsted, 1964; FAO/WHO, 1965).

A number of investigators have questioned the validity of the value of 2 mg of nitrogen/basal kcal for the relationship between basal metabolic rate and endogeneous urinary nitrogen excretion. Table 1 summarizes the results of studies designed specifically to evaluate endogeneous urinary nitrogen losses in man. These studies were conducted under a variety of conditions over a span of more than 45 yr. The average of all measurements is approximately 1.3 mg of nitrogen/basal kcal, with a coefficient of variation of about 20 percent. All values are well below the 2-mg figure proposed originally by Smuts on the basis of animal studies.

Although information on children and infants is limited, it provides additional support for the view that the endogenous urinary nitrogen excretion of human subjects is less than 2 mg/basal kcal. Children averaged 1.1 mg, and infants 0.7 mg, of N/basal kcal. As pointed out by Fomon *et al.* (1965), Ashworth's (1935) studies indicated that urinary excretion of endogenous nitrogen by weanling rats was 1 mg/basal kcal and by adult rats, 1.5 mg/basal kcal. Furthermore, Ashworth and Cowgill (1938) demonstrated that the ratio of endogenous nitrogen excreted in the urine to basal caloric expenditure increased rapidly during the period of rapid growth after weaning.

On the basis of the values for endogenous nitrogen excretion shown in Table 1, it appears that a sex difference may exist. Consequently, the averages are listed separately for the adult males and females. The significance of the difference has not been evaluated, as but few experiments have been conducted in which males and females were studied under the same conditions. In one of these (Hawley *et al.*, 1948) a slightly higher value for males was obtained but the difference was not statistically significant.

Accurate estimation of minimum endogenous urinary nitrogen excretion is difficult. The output of subjects fed a protein-free diet declines rapidly for a few days and then tends to stabilize (Martin and Robinson, 1922; Munro, 1964). Because this curve lacks a sharp inflection point, the relatively constant value obtained during the period when nitrogen output plateaus, usually after 5–10 days of nitrogen depletion, is taken as the minimum. Estimates made before depletion may be more realistic, since the body is known to adapt to low intakes of nitrogen (Waterlow, 1968); in fact, Mitchell (1964) and Holmes (1965) have concluded that the lower values reported for endogenous nitrogen loss are unrealistic. Nevertheless, selection of an appropriate point during the period of declining nitrogen loss becomes arbitrary.

Metabolic Fecal Nitrogen Excretion

The results of studies in which fecal nitrogen losses of subjects consuming no protein were estimated are summarized in Table 1. Unlike the values for endogenous urinary nitrogen, they show a variation from the average of more than 80 percent. Undoubtedly, a significant component of such variation is due to the difficulty of assuring proper assignment of collections to respective metabolic periods and complete fecal collection. Also, fecal nitrogen excretion is influenced by the quantity and composition of the food consumed (Mitchell, 1964).

Values for infants and children are lower in g/day, but represent a higher percentage of the total nitrogen loss (approximately one-third) than that for adults (about 27 percent). The average value of 0.85 g of metabolic fecal nitrogen daily for a 70-kg man appears to be realistic. This is about 12 mg/kg of body weight.

Integumental Nitrogen Loss

The values in Table 1 summarize measurements of skin losses of nitrogen for primarily sedentary individuals in temperate climatic regions. The higher dermal losses in males due to beard growth and sweating may be balanced in the female by losses during menstruation (Mitchell and Edman, 1962). The average value of 300 mg/day for dermal losses has been considered sufficient to include the nitrogen required for hair and nail replacement and probably for other minor nitrogen losses (Calloway et al., 1971).

Estimate of Total Nitrogen Loss

Estimates of endogenous, metabolic, and integumental nitrogen losses would provide a basis for estimating the minimum amount of dietary nitrogen needed if efficiency of nitrogen utilization were 100 percent. The total nitrogen loss equals the endogenous urinary (*UN*) plus the metabolic fecal (*FN*) plus the integumental (*IN*) nitrogen losses:

$$\text{Total nitrogen loss} = UN + FN + IN.$$

Utilizing the values estimated above, average adult nitrogen loss can be calculated for the RDA (recommended dietary allowance) reference man (FNB, 1968) as follows: Basal energy requirement for the 70-kg reference man is 1,750 (1,750 ÷ 70 = 25 kcal/kg). Endogenous urinary nitrogen loss is estimated to average 1.3 mg/basal kcal. Therefore, 1,750 × 1.3 = 2.28 g of *UN*/day. Thus: 2.28 + 0.85 (*FN*) + 0.30 (*IN*) = 3.43 g of

TABLE 1 Endogenous Urinary, Metabolic Fecal, and Integumental Nitrogen Losses of Human Subjects[a]

Urine			Feces			Integument		
Ref.[c]	No. of Subjects and Sex	Average (mg/basal Kcal)	Ref.[c]	No. of Subjects and Sex	Average (g/day)	Ref.[c]	No. of Subjects and Sex	Average (g/day)
a	A 2 M	1.5	a	A 2 M	1.10	l		0.333
b	A 1 M	1.7	b	A 1 M	0.73	m		0.340
c	A 1 M	1.4	c	A 2 F	0.53	n		0.372
d	A 9 F	1.4	d	A 48 F-M	1.07	o	A 4 M	0.360
e	A 5 M	1.4	e	A 4 M	0.95	p	A 8 M	0.254
f	A 4 M	1.5	f	A 13 F-M	0.88	q	A 20 M	0.143
			r	A 4 M	1.07			
g	A 6 F	1.2	h	A 8 M	0.66			
g	A 7 M	1.3						
r	A 4 M	1.4	k	A 6 F-M	0.40			
h	A 8 M	1.6	s	A 25 F	0.50			
s	A 25 F	1.1						

			Mean 1.3[b]				Mean 0.85[b]		Mean 0.300
			Female, 1.2						
			Male, 1.5[b]						

i,j	C 12 F-M	0.9	i,j	C 12 F-M	0.36		
	C 9 F-M	1.1		C 9 F-M	0.45		
	C 7 F-M	1.4		C 7 F-M	0.51		
Mean		1.1[b]	Mean		0.43[b]		
j	I 6 F	0.6	i	I 7 F	0.15		
	I 4 M	0.8		I 12 M	0.27		
Mean		0.7[b]	Mean		0.23[b]		

[a] A = Adult (17 yr +); C = Child; I = Infant (under 1 yr).

[b] Weighted averages.

[c] References: a (Martin and Robinson, 1922); b (Smith, 1926); c (Deuel et al., 1928); d (Bricker et al., 1945); e (Murlin et al., 1946); f (Mueller and Cox, 1947); g (Hawley et al., 1948); h (Young and Scrimshaw, 1968); i (Fomon et al., 1965); j (DeMaeyer and Vanderborght, 1961); k (Hegsted et al., 1946); l (Voit, 1930); m (Cuthbertson and Guthrie, 1934); n (Freyberg and Grant, 1937); o (Mitchell and Hamilton, 1949); p (Darke, 1960); q (Sirbu et al., 1967); r (Gopalan and Narasinga Rao, 1966); and s (Bricker and Smith, 1951).

N/day or 0.049 g/kg of body wt/day. This is equivalent to 21.4 g
(3.43 × 6.25) per day of protein with 100 percent efficiency of utilization or 0.31 g protein/kg of body weight (21.4 ÷ 70). The same procedure applied to the reference woman (58 × 0.31) yields 18 g of completely utilizable protein or (58 × 0.049) 2.84 g of N/day.

In a paper published after Table 1 was compiled, Calloway and
Margen (1971) reported the mean daily endogenous urinary nitrogen
loss for 13 men with an average daily basal metabolic rate of 1,875 kcal
to be 2.41 g or 1.3 mg (2,410 ÷ 1,875) per basal kcal, the same as the
average value in Table 1. The mean daily fecal nitrogen of 0.96 g found
in their study was somewhat above the average value of 0.85 g shown in
Table 1, but is within the range of values reported. The average cutaneous nitrogen loss of 129 mg/day is included in Table 1, as these subjects
represent a segment of the 20 male subjects in one of the studies cited
(Sirbu *et al.*, 1967). Finally, the average daily nitrogen loss in their study
was 2.41 (*UN*) + 0.96 (*FN*) + 0.129 (*IN*) = 3.5 g, which, divided by the
average body weight of the subjects (70.8 kg), is 0.049 g/kg of body
weight, the same as the value obtained from the information summarized
in Table 1.

The value of 49 mg of N/kg of body weight for daily nitrogen loss is
just over half the amount proposed by the FAO/WHO (1965) expert
committee. That committee accepted the value of 2 mg/basal kcal for
endogenous urinary losses for animals as appropriate for man, overestimated fecal losses, and equated the requirement for "adult growth"
(discussed below) proposed by Mitchell (1964) with dermal and minor
nitrogen losses. The estimates for these components were therefore substantially in excess of those obtained by direct measurement and resulted in overestimation of total nitrogen loss.

Several reports (Sherman *et al.*, 1920; Sumner and Murlin, 1938;
Bricker *et al.*, 1945; Hegsted *et al.*, 1946; Calloway and Margen, 1971)
indicate that such high-quality proteins as those of milk and egg, as well
as the proteins of mixed diets, are used for maintenance by human
adults with only 60–70 percent efficiency. These observations suggest
that if the value of 49 mg of N/kg of body weight is increased by about
35 percent, the resulting value of 75 mg/kg should approximate the
dietary nitrogen requirement for maintenance of human adults.

THE NITROGEN BALANCE APPROACH

Requirement for Nitrogen Equilibrium in the Adult

Measurement of the least amount of protein that will maintain human subjects in nitrogen equilibrium has been widely used over the years to estimate nitrogen and protein requirements of adults. According to Terroine (in Waterlow and Stephen, 1957), this is the most valid procedure.

Nasset (1956) has emphasized that the results of balance studies can be influenced by many factors, e.g., environment, previous diet, total energy and nitrogen intake, and the time span and composition of the experimental diet. Wallace (1959) has pointed out the inherent cumulative errors of the balance technique: "Even with the most refined and meticulous technique a finite quantity of the measured intake is lost in the process of feeding and, similarly, a finite portion of the excreta is not recovered. When output is subtracted from intake, the two losses are additive, not self-cancelling as is generally supposed. The problem may be stated in another way: the numerical value obtained for intake is always larger than actuality while the numerical value for output is always smaller. Subtraction of the two numerical values compounds the error." The extent to which requirements may be overestimated by the nitrogen balance procedure as a consequence of the difficulty of identifying a clear end point has not been established. When the point at which the requirement is met must be detected by distinguishing among a series of small differences that are determined by subtracting a large value for intake from a large value for excretion, the potential for error is great. The wide variation among individuals (coefficient of variation, 15–20 percent) introduces another potential source of error whenever experiments can be done with only a few subjects at a time. Effects of adaptation to differing nitrogen intakes present still another problem. And the probability that a subject in equilibrium will be in positive balance one day and negative balance the next poses yet another.

Not uncommonly in balance studies, efforts to relate apparent changes in nitrogen retention to changes in body composition have shown that the estimates of retained nitrogen were impossibly high. However, average values for nitrogen retention of adults fed amounts of protein in the requirement range are more in accord with those predicted, especially if allowance is made for dermal and other losses that are rarely measured and for the additive errors of the nitrogen balance procedure. Although the pitfalls of balance studies are recognized by most of those using the balance technique, inherent errors and dermal losses are seldom considered in assessing results. Hegsted (1963) has used a positive nitrogen balance of 0.5 g/day as the criterion for attaining nitrogen equilibrium.

Sherman *et al.* (1920) reviewed the results from 25 different investigations of 109 nitrogen balance experiments on men and women who were fed a wide variety of protein sources. They concluded that the average nitrogen requirement was 96 mg/kg of body wt/day, nearly double the value of 49 mg/kg for the estimated daily nitrogen losses. These results imply that efficiency of nitrogen utilization, when nitrogen intake met the requirement, was just over 50 percent. This is low compared to information from subsequent studies, even with mixed diets; and the requirement value is somewhat high, but both are within the range of values reported. Lusk (1920) concluded from reviewing early nitrogen balance experiments that adult man could be maintained in good health and nitrogen equilibrium with an intake of somewhat above 4 g and somewhat below 6 g of nitrogen daily, equivalent to 57–86 mg/kg/day or 25–40 g of protein per day for a 70-kg man.

Information from two subsequent studies designed to evaluate proteins as nitrogen and amino acid sources for maintenance of nitrogen equilibrium in human adults is summarized in Table 2. Bricker *et al.* (1945) tested several single sources of protein, whereas Hegsted *et al.* (1946) used combinations simulating a mixed dietary regimen. The values obtained ranged from 62 mg of N/kg for milk protein to 106 mg of N/kg for white flour protein. The amounts of protein nitrogen required to maintain adults in nitrogen equilibrium were estimated from regression analysis. They ranged from 0.065 to 0.085 g/kg of body wt/day, with an average of 0.075. Values obtained with protein sources such as milk, meat–vegetable mixtures, and soybean flour were very

TABLE 2   Protein Nitrogen Needed by Adults for Nitrogen Equilibrium or Slight Positive Balance

Ref.	No. of Subjects and Sex	Protein Nitrogen Source	Nitrogen Required for Equilibrium (g/day)	(g/kg)	Nitrogen Utilization[a] (percent)
Bricker	4 or 5 F	Milk	3.59	0.062	70
et al.,	4 or 5 F	White flour	6.19	0.106	40
1945	4 or 5 F	Soybean flour	3.74	0.064	60
	4 or 5 F	Soybean–white flour[b]	4.39	0.076	52
	4 or 5 F	Mixed foods	4.06	0.070	58
Hegsted	26 M–F	Veg. mixture	4.99	0.076	64
et al.,	16 M–F	+ Meat[c]	4.06	0.064	72
1946	6 M–F	+ Bread[c]	4.99	0.083	64
	6 M–F	+ Soya flour [c]	4.99	0.082	59
	6 M–F	+ Wheat germ[c]	4.06	0.067	76
Mean				0.075	

[a]Digestibility X biological value.
[b]Soybean protein, 36 percent; white flour protein, 64 percent.
[c]Substituted on an isonitrogenous basis.

similar and are all higher than the estimated daily nitrogen loss of 49 mg/kg. The average intake of N from these protein sources (75 mg/kg/day) that is required to maintain equilibrium is almost identical to that reported recently for egg proteins (Calloway and Margen, 1971). Although miscellaneous nitrogen losses were not taken into account in calculating the values in Table 2, trials in which efficiency of nitrogen utilization was low were included in calculating the average. Since the average of the four lowest values is only 65 mg of N/kg of body wt/day, a value of 75 mg should allow amply for unmeasured nitrogen losses when the dietary protein is of high quality. This would be equivalent to 0.47 g of protein per kg of body wt/day (0.075 × 6.25), or 27 g (0.47 × 58) and 33 g (0.47 × 70) of protein daily for the reference adult female and male, respectively.

Obviously, none of the proteins tested was used with an efficiency approaching 100 percent. The value for milk proteins, for example, was only 70 percent; and the lowest value, that for white flour proteins, 40 percent. Also, the differences among proteins as nitrogen sources were not as great as might be predicted from comparison of Net Protein Utilization (NPU) values. Values below the maximum for efficiency of nitrogen utilization are to be expected in these experiments, however; because estimates of nutritional value, routinely made under conditions in which protein is limiting, tend to fall as the protein content of the diet approaches adequacy (Mitchell, 1924; Miller and Payne, 1969). Mitchell (1948) has emphasized that measurements of Biological Value (BV) in adult man yield values that are low for high-quality proteins and that exhibit a discouraging lack of agreement among different laboratories. Moreover, evidence indicates that high-quality proteins may not be used with maximum efficiency by adults even when protein intakes are less than adequate (Sumner and Murlin, 1938) (see also D. M. Hegsted, pp. 64–88 in this volume).

Bricker *et al.* (1949) tested the adequacy of their estimate of the requirement for protein nitrogen in nine female adults by observing them for a period of 10 weeks while they consumed a diet that provided an amount of nitrogen equal to their previously determined requirements for nitrogen equilibrium, plus an allowance for "adult growth." The estimated average daily protein nitrogen requirement of 5.08 g represented the sum of 3.80 g for nitrogen equilibrium and 1.28 g for "adult growth." With this level of intake, the subjects remained in positive nitrogen balance of about 0.6 g/day throughout the experiment. At the end of the 10 week period, there was no tendency for balance to be less positive, no change in hemoglobin or erythrocyte count, and no evidence of deterioration in several performance tests.

The term "adult growth" is derived from observations on the apparent storage of nitrogen by adults in long-term experiments, such as that of Grindley and Mitchell (1917), in which the average positive nitrogen retention, after making allowance for 400 mg/day for dermal losses, was about 1 g of N/day for 220 days. The term, and its use for estimating adult protein requirements, has been controversial (Smuts, 1935; Hegsted, 1964). In view of the fact that it represents nearly one-quarter of the estimated maintenance requirement, this is not surprising. Bricker *et al.* (1949) predicted a daily nitrogen loss of 2.94 g (0.049 X 60). This value, after adjusting for the 61.4 percent efficiency of protein utilization observed, would give an estimated daily requirement of 4.8 g (2.94 ÷ 0.614). Considering that a positive balance of about 0.5 g/day is probably required to ensure nitrogen equilibrium, the intake by subjects over an experimental period of 10 weeks of 5.08 g of nitrogen, which produced an average daily positive nitrogen balance of 0.61 g, was just 6 percent greater than the estimated requirement, without allowance for "adult growth." Underestimation of endogenous urinary nitrogen loss by only 5 mg/kg would give a value of 5.1 g of nitrogen for the requirement of these subjects. The concept of "adult growth" is, therefore, unrealistic; retention of 1 g of N/day (Bricker *et al.*, 1945, 1949) should be accompanied by an increase in body weight of about 2 kg/yr.

In studies of nitrogen retention in men fed amino acid diets, Rose and Wixom (1955c) concluded that equilibrium could be maintained with a daily nitrogen intake of 3.5 g. This is little more than the estimate of nitrogen loss obtained using the factorial method and is often quoted to support the validity of that method. However, the two subjects in the study had positive nitrogen balances of only 0.15 g/day and required intakes of 4–6 g of nitrogen to achieve positive balances of 0.26–0.46 g/day with energy intakes of 54 or 55 kcal/kg of body weight. Also, after consuming low-nitrogen diets for 20 days or more, they had adjusted to lower intakes. Since no allowance was made for integumental losses in this study, and assuming small positive errors in the nitrogen balance technique, a requirement between 4–6 g/day seems more realistic than the 3.5 g proposed and is in accord with the results of studies in which proteins have been fed. This suggests that even a diet in which the amounts of all of the essential amino acids have been adjusted to resemble the amino acid requirements tends to be used inefficiently as the nitrogen requirement is approached. In fact, efficiency in retention of the nitrogen in excess of 3.5 g to give a positive balance of 0.46 g/day was only about 15 percent.

Protein Nitrogen versus Total Nitrogen

The studies summarized in Table 3 involved adults fed primarily diets containing a single food source of protein, together with nonspecific nitrogen from simple chemical compounds. The results imply that, if

TABLE 3   Amounts of Protein Nitrogen that Maintained Nitrogen Equilibrium or Slightly Negative Balance in Adults

Ref.	No. of Subjects and Sex	Protein Source	Nonspecific Nitrogen Supplement	Total Nitrogen Intake (g/day)	Intake of Protein Nitrogen (g/day)	(g/kg)
a	3 F	Eggs[a]	DAA[b]	6.5	1.30	0.023
a	1 F	Eggs[a]	DAA[b]	6.5	1.60	0.029
a	2 F	Eggs[a]	DAA[b]	10.0	1.30	0.023
a	1 F	Eggs[a]	DAA[b]	10.0	1.60	0.029
b	1 M	Eggs[a]	Gly and DAC[c]	6.5	1.60	0.025
b	1 M, 4 F	Eggs[a]	Gly and DAC[c]	6.5	2.00	0.035
b	1 M	Eggs[a]	Gly and DAC[c]	6.5	2.40	0.032
b	2 M, 1 F	Eggs[a]	Gly and DAC[c]	13.0	1.60	0.027
b	2 M, 4 F	Eggs[a]	Gly and DAC[c]	13.0	2.00	0.034
b	1 M, 1 F	Eggs[a]	Gly and DAC[c]	13.0	2.50	0.043
b	2 M, 2 F	Eggs[a]	Gly and DAC[c]	13.0	3.00	0.052
c	4 M, 2 F	Rice	None	6.6	6.00	0.085
c	5 M, 1 F	Rice	Gly and DAC	12.7	5.00	0.061
d	10 M	Corn	None	7.7	7.00	0.061
e	10 M	Corn	Gly and DAC	8.0	6.00	0.079
f	3 M	Corn	Gly and DAC	12.5	4.00	0.048
f	4 M	Corn	Gly and DAC	12.5	4.50	0.058
f	2 M	Corn	Gly and DAC	12.5	5.00	0.063
g	8 M	Eggs[d]	None	4.45	4.45	0.062
g	6 M	Eggs	Gly and DAC	4.45	3.12	0.043
g	5 M	Eggs	Gly and DAC	4.45	2.67	0.037
h	9 M	Beef[d]	None	4.46	4.46	0.062
h	5 M	Beef	Gly and DAC	4.46	3.66	0.050
h	3 M	Beef	Gly and DAC	4.46	3.34	0.046
h	2 M	Beef	Gly and DAC	4.46	3.12	0.043
i	14 M	Cow's Milk[d]	None	4.31	4.30	0.061
i	7 M	Cow's Milk	Gly and DAC	4.31	3.44	0.048
i	4 M	Cow's Milk	Gly and DAC	4.31	3.23	0.045
i	10 M	Cow's Milk	DAA	4.70	3.54	0.046

[a]The basal diet provided 0.5 g of nitrogen daily from fruits and vegetables.
[b]Dispensable amino acids.
[c]Glycine and diammonium citrate.
[d]Most of the subjects in these studies were in slight negative nitrogen balance. The protein source represented 90 percent of the total nitrogen.
[e]References: a (Swendseid et al., 1960); b (Swendseid et al., 1959); c (Chen et al., 1967); d (Kies et al., 1965a); e (Kies et al., 1965b); f (Kies et al., 1967); g (Scrimshaw et al., 1966); h (Huang et al., 1966); and i (Scrimshaw et al., 1969).

total nitrogen intake is 6.5 g/day or more, part of the high-quality protein can be replaced by dispensable amino acids or such nonspecific sources of nitrogen as glycine and diammonium citrate and that the amount of nitrogen from intact protein needed to maintain nitrogen equilibrium may fall to as low as 0.03 g/kg/day, representing less than 0.2 g of protein/kg/day.

Experiments by Swendseid *et al.* (1959, 1960) support the generally accepted view that egg protein is a superior dietary source of amino acids for man. Their observations suggested that, when the quantity of egg protein in the diet is reduced, total nitrogen becomes limiting before any of the essential amino acids do so. The concept that nitrogen may become limiting in high-quality proteins before specific amino acids do was proposed by Snyderman *et al.* (1962), who observed that the growth of young infants fed a cow's milk formula in which the protein was diluted with nonspecific sources of nitrogen was satisfactory. For two-thirds of the subjects examined by Swendseid *et al.,* less than 3 g/day of egg protein nitrogen met the need for specific nitrogen compounds, provided that the total nitrogen intake was at least 6.5 g. As a source of nonspecific nitrogen, the dispensable amino acids tended to be superior to a mixture of glycine and diammonium citrate.

The concept that total nitrogen may become limiting before essential amino acids do, when adults are fed small amounts of high-quality proteins, also received some support from studies on young men by Scrimshaw and associates (1966, 1969) and Huang *et al.* (1966). In one study (Scrimshaw *et al.*, 1966) most of the subjects, with intakes of about 4.5 g of N/day, when egg protein provided 90 percent of the total nitrogen, were not quite in nitrogen equilibrium; when the proportion of total nitrogen supplied by egg protein was decreased to 60 percent, i.e., less than 3 g/day, and total nitrogen was maintained at 4.5 g by substituting glycine and diammonium citrate for part of the protein, nitrogen balance was essentially unaltered (Table 3). Results obtained in studies of men fed cow's milk or beef protein (Huang *et al.*, 1966; Scrimshaw *et al.*, 1969) were essentially similar to those obtained with eggs.

With such low-quality proteins as those of wheat, rice, and corn, the amount of nitrogen required to maintain nitrogen equilibrium in adult man is between 6 and 8 g/day (Chen *et al.*, 1967; Kies *et al.*, 1965a,b, 1967)—more than the amounts required from high-quality proteins. Yet with these too, when as much as 25 percent of the intact protein was replaced by nonspecific sources of nitrogen, nitrogen balance appeared not to be impaired.

These observations pose the question as to whether requirements for essential amino acids are influenced by total nitrogen intake. It is evident that our understanding of relationships between total nitrogen intake and specific amino acid needs is incomplete and merits further study. It may be that, as the requirement for total nitrogen is approached, efficiency of utilization of amino acids for tissue protein synthesis falls and at different rates for different amino acids or, conversely, that the extent of destruction of amino acids increases with increasing protein intake, but at different rates for different amino acids. Metabolic adaptations, such as more efficient reutilization of amino acids from the body pools as the intake of an amino acid falls (Harper, 1965; Waterlow and Stephen, 1969), may be responsible for some of these effects; this possibility deserves attention. Until these relationships are better understood and more quantitative information about factors that influence amino acid requirements has accumulated, an adequate explanation for these observations will not be possible.

REQUIREMENTS OF INFANTS AND CHILDREN

Protein requirements of infants are estimated by measuring the amount of nitrogen from milk or prepared formulas that will support satisfactory weight gain. Protein needs of infants fall rapidly during the first year of life (Hegsted, 1957). Fomon (1959, 1960, 1961a), Fomon and May (1958), Fomon and Filer (1967); and Fomon et al. (1969) have reported protein intakes of infants fed either a cow's milk formula or human milk during the first 6 months of life or a soybean formula at age 5–6 months. The diets supported satisfactory rates of growth. Average values, calculated from their reports, as summarized in Table 4, indicate that whether human milk or cow's milk is the source of protein, the requirement falls from about 2.2 g/kg/day shortly after birth to 1.5 g/kg/day by 6 months of age. This is in agreement with earlier observations of Beach et al. (1941). Both Fomon (1967) and Chan and Waterlow (1966) conclude that, by 1 year of age, protein requirement has fallen further to 1.0–1.25 g/kg/day.

Many other studies have been conducted on infants fed much higher intakes of protein than those reported in Table 4. Fomon (1961b) has summarized several of these; and he, Hegsted (1957), and Woodruff (1961) have discussed them. Evidence that nitrogen retention is higher when intakes are higher than those reported in Table 4 has not been accompanied by evidence of greater growth rates. Calculations of body composition based on such nitrogen retention lead to values for body protein content that are unrealistically high, suggesting, as Wallace

TABLE 4   Average Nitrogen and Protein Intakes of Infants with Satisfactory
Growth Rates Fed Cow's Milk Formula, Human Milk, or Soybean Formula

Age (mo)	Human Milk[a]		Cow's Milk Formula[b]		Soybean Formula[c]	
	Nitrogen (mg/kg/day)	Protein (g/kg/day)	Nitrogen (mg/kg/day)	Protein (g/kg/day)	Nitrogen (mg/kg/day)	Protein (g/kg/day)
0.5	387	2.4	311	2.0		
1.5	351	2.2	288	1.8		
2.5	282	1.8	261	1.6		
3.5	280	1.7	239	1.5		
4.5	255	1.6	243	1.5	298	1.8
5.5	235	1.5	238	1.5	281	1.7

[a]Fomon and May, 1958.
[b]Fomon, 1960; Fomon *et al.*, 1969.
[c]Fomon, 1959.

(1959) has emphasized, that nitrogen balance studies are less adequate
as a method of determining nitrogen requirements of infants than are
growth studies.

The requirement for protein (N × 6.25) falls gradually after 1 yr until
it reaches the adult value of about 0.5 g/kg of body wt/day at maturity.
Estimates of the rate of fall by Hegsted (1957) and Waterlow (1970)
suggest the following pattern:

Age	9 mo	15 mo	2 yr	6 yr	12 yr	15 yr	adult
g/kg	1.25	1.0	0.9	0.8	0.7	0.6	0.5

Information on the maintenance requirements of children and infants is
limited. In view of the growth demands and the difficulty of experimentally separating maintenance and growth requirements, this is not surprising. DeMaeyer and Vanderborght (1961) have calculated the nitrogen requirement for maintenance of children aged 3–7 yr from
regression equations. Values of 0.08–0.1 g of N/kg of body weight
(equal to 0.5–0.625 g of protein) were obtained with proteins of
excellent-to-good quality. Also, the amounts consumed from cow's
milk by infants between 1 and 2 months of age who failed to grow
(Snyderman *et al.*, 1962)—0.14 to 0.26 g of N/kg/day (0.88–1.6 g of
protein)—might be taken as an indicator of maintenance needs (Table 5).

The studies of Snyderman *et al.* (1962) with 3-week- to 3-month-old
infants (Table 5) are unique. Only four subjects were involved, but the
results suggest that, when the quantity of cow's milk in the diet is decreased, total nitrogen may become limiting before the essential amino
acids do. The infants gained weight normally with intakes of 0.28–0.49 g

TABLE 5 Protein Nitrogen Needed for Growth (Normal Gain) and Maintenance (No Gain) of Infants Fed Cow's Milk Diluted with Nonspecific Nitrogen Sources[a]

Subjects' Sex and Age	Total Nitrogen Intake (g/kg)	Nonspecific Nitrogen Supplement	Intake of Protein Nitrogen	
			Normal Gain (g/kg)	No Gain (g/kg)
F, 1 mo	0.42	None	0.42	—
	0.26	None	—	0.26
	0.80	Gly	0.21	—
	0.80	Urea	0.21	—
M, 3 mo	0.31	None	0.31	—
	0.24	None	—	0.24
	0.75	Gly or Urea	0.20	—
M, 3 wk	0.49	None	0.49	—
	0.14 or 0.69	None or Urea	—	0.14
	0.69	Urea	0.18	—
M, 3 wk	0.28	None	0.28	—
	0.18	None	—	0.18
	0.68	Urea	0.18	—

[a]Snyderman et al., 1962.

of N/kg of body wt/day from cow's milk (1.8–3.0 g of protein). These compare with values of 0.26–0.35 g/kg body wt/day (1.6–2.2 g of protein) for infants of comparable age fed human milk or cow's milk formulas (Table 4). However, when the latter formula was modified and the total nitrogen content was augmented with glycine or urea, 0.18–0.21 g of N/kg of body wt/day (1.1–1.3 g of protein) provided enough of the essential amino acids for normal gain. It is difficult to determine accurately how much the requirement of these infants for protein was decreased by the procedure, but the results would suggest that on the order of 30 percent of milk protein can be replaced by nonspecific sources of nitrogen. This is less, as would be anticipated, than was observed with adults in the studies of Swendseid et al. (1959, 1960).

AMINO ACID REQUIREMENTS

Requirements Determined with Amino Acid Diets

Since 1954, when Rose and associates (1954) identified the amino acids that were essential for man, researchers have attempted to estimate quantitative requirements for infants, children, women, and men fed crystalline amino acids in diets in which the amount of each acid could be varied independently of all of the others. Although food proteins are

TABLE 6 Amino Acid Requirements of Man (mg/kg/day)

Amino acid	Infants (2–6 mo)[a]	Children (10–12 yr; 36 kg)[b]		Adults[c,d,e] Female (58 kg)				Male (70 kg)	
	mg/kg	mg/day	mg/kg	mg/day	mg/kg	mg/day	mg/kg	mg/day	mg/kg
His	34	?	?	?	?	?	?	?	?
Ile	119	1,000	28	450 (250–450)	7.8	550	9.5	700 (650–700)	10.1
Leu	150	1,500	42	620 (170–620)	10.7	725	12.5	1,100 (500–1,100)	15.7
Lys	103	1,600	44	500 (400–500)	8.6	545	9.4	800 (400–800)	11.4
Met	45	800	22	700[f] (300–700)	12.1	700	12.1	1,100 (800–1,100)	15.7
Cys	80	–	–	–	–	–	–	–	–
TSAA[j]	*125*	*800*	*22*	*700*	*12.1*	*700*	*12.1*	*1,010[g] (910–1,010)*	*14.4*
Phe	90	800	22	700[h] (600–700)	12.1	700	12.1	1,100 (800–1,100)	15.7
Tyr	present	–	–	–	–	–	–	–	–
TAAA[j]	*?*	*800*	*22*	*700*	*12.1*	*700*	*12.1*	*1,100[i]*	*15.7*
Thr	87	1,000	28	305 (103–305)	5.3	375	6.5	500 (300–500)	7.1
Trp	22	120	3.3	160 (82–157)	2.8	168	2.9	250 (150–250)	3.6
Val	105	900	25	650 (465–650)	11.2	622	10.7	800 (400–800)	11.4
Total	**835**	–	**288**	–	–	–	**76**	–	**91**

[a] Holt and Snyderman, 1965.
[b] Nakagawa et al., 1960, 1961a,b, 1962, 1963, 1964.
[c] Leverton, 1959.
[d] Hegsted, 1963.
[e] Rose, 1957.
[f] Reynolds et al., 1958.
[g] Rose and Wixom, 1955c. Cys will spare 80–89 percent of the Met requirement.
[h] Burrill and Schuck, 1964.
[i] Rose and Wixom, 1955c. Tyr will spare 70–75 percent of the Phe requirement.
[j] TSAA is total sulfur-containing amino acids; TAAA is total aromatic amino acids.

the only practical dietary source of nitrogen, amino acid diets will support growth and maintenance of man and other animals. Originally, casein proved superior to hydrolyzed casein or mixtures of amino acids in supporting nitrogen equilibrium in adult man (Rose, 1957); additional energy appeared to be required when amino acids were the nitrogen source. Recent observations (Anderson *et al.*, 1969) have not confirmed the general implications of the earlier studies. They do suggest, however, that more energy may be required to maintain nitrogen equilibrium with diets containing only amino acids than with those containing intact protein, whenever a large proportion of the nitrogen in the amino acid diets is from such sources as glycine, diammonium citrate, or urea.

Estimated amino acid requirements have been summarized by Holt and Snyderman (1965) for infants, by Nakagawa *et al.* (1964) for children, by Leverton (1959) for women, and by Rose (1957) for men. In addition, Hegsted (1963) has reviewed the information on requirements of adults. He plotted all of the values available at that time to indicate the degree of variability; and, using the values for women obtained by Leverton and some reported by others for several of the amino acids, estimated requirements of regression analysis. This information is summarized in Table 6. The values for children are averages calculated from the information on the individual subjects given in the published reports of Nakagawa *et al.* (1960, 1961a,b, 1962, 1963, 1964). The values for children, women, and men originally reported as grams per day have also been expressed per kilogram of body weight.

Histidine is listed as essential only for infants (Snyderman *et al.*, 1963), as no quantitative requirement has been established for other age groups (Rose, 1957), even though its dispensability for adult man has been questioned (Nasset and Gatewood, 1954; Anonymous, 1964).

Values for the sulfur-containing amino acids (SAA) and the aromatic amino acids (AAA) pose a special problem. Rose and Wixom (1955a,b) concluded that, for adult man, 80–89 percent of the required methionine could be replaced by cystine and 70–75 percent of the required phenylalanine could be replaced by tyrosine. A question, therefore, arises about the effects of including different proportions of these pairs of amino acids in diets used for studying their requirements. The requirement of men (Rose and Wixom, 1955a) for the sulfur-containing amino acids was less when both methionine and cystine were included than when the diet contained only the former. The sparing effects of cystine and tyrosine are less in the young rat than in the adult, but neither the effect of cystine on the methionine requirement nor that of tyrosine on the phenylalanine requirement of the human infant has been established.

There appears to be a sex difference in the amino acid requirements of adults. However, when the requirements of men and women are expressed per kilogram of body weight and when the revised estimates of Hegsted (1963) for women are compared with those for men, the differences are not great. Rose and associates accepted as the requirement the highest value for any individual man, whereas the values arrived at by Hegsted for women have been calculated, by regression analysis, as the amounts needed to maintain nitrogen equilibrium; this may account for the higher values for men. In view of the marked variability of the requirements (Hegsted, 1963) and the difference in the criteria of estimation, it seems unlikely that there is a true sex difference.

The most striking feature of amino acid requirements is how low the values are for adults. The total essential amino acid requirements of infants represent a little more than 40 percent of the protein requirement of about 2 g/kg/day; for children, around 36 percent of about 0.8 g/kg/day. The proportion of total essential amino acids required by infants is not far below the proportion in high-quality proteins and may exceed those in some cereal proteins. In contrast, comparison of the total essential amino acid requirements of adults (average of values for men and women) with the protein requirement of 0.47 g/kg/day indicates that only some 20 percent of the required nitrogen need comes from essential amino acids. These calculations would suggest, as has been observed, that high-quality proteins can be diluted considerably with nonspecific nitrogen sources without impairing nitrogen retention—provided total N requirements are met.

The possibility that the estimated amino acid requirements of adults are low must also be considered. Most investigators, when they estimate the requirement for a given amino acid, include all others in excess in the diet. If analogy to the law of mass action is valid, this procedure should ensure high efficiency of utilization of the one that limits protein synthesis. A similar procedure, when used in studies of amino acid imbalances, increases the incorporation of the limiting amino acid into liver proteins (Yoshida *et al.*, 1966; Benevenga *et al.*, 1968; Soliman and King, 1969). The use of maintenance of nitrogen equilibrium as the criterion for meeting the requirement may also lead to an underestimate, in view of the need for slight positive balance to ensure that equilibrium has been achieved.

Amino Acid Intakes of Adults Consuming Low-protein Diets

Amino acid requirements can also be estimated by calculating, from the amino acid composition of proteins (Orr and Watt, 1968; FAO, 1970),

the amounts of amino acids consumed when protein intake is just sufficient to maintain nitrogen equilibrium in the adult or to support satisfactory growth of the infant or child (Harte and Travers, 1947).

The information on amino acid intakes assembled in Table 7 was calculated from the results of three of the experiments reported in Table
2, in which single protein sources were used for estimation of the protein requirements of women (Bricker et al., 1945); two of the studies
reported in Table 3, in which egg protein was fed at extremely low
levels while total nitrogen intake was kept high (Swendseid et al., 1959,
1960); two similar experiments with cereal grains as the major source of
protein (Chen et al., 1967; Kies et al., 1967); two groups from the experiments (Table 3) in which milk protein was fed at a suboptimal level
of intake and the diet was further diluted with a source of nonspecific
nitrogen (Scrimshaw et al., 1969); and one in which a combination of
flour and milk solids was fed, with about 10 percent of the protein provided by the latter (Goyal and Clark, 1964). Amino acid intakes calculated from the results of other experiments were similar to or higher
than the values included in Table 7.

Swendseid et al. (1960) obtained a positive nitrogen balance of 0.31 g
in eight trials on young women consuming 1.3 g of N from egg, with
total nitrogen intakes of 6.5 or 10 g/day, largely from nonspecific N
sources. The amino acid intakes of these subjects, from the egg protein
(Columns 4 and 12, Table 7) resemble closely the values for requirements calculated by Hegsted (1963) (Column 7, Table 6), except for
the sulfur-containing amino acids and possibly tryptophan, which are
low. It should be noted that the basal diet included 0.5 g of nitrogen
from fruits and vegetables, the composition of which is unknown; the
calculated amino acid intakes could be 20–30 percent low. The values
for men in this investigation (Columns 5 and 11) were calculated from
intakes for only two subjects, with nitrogen retentions averaging 0.68
g/day (Swendseid et al., 1959). Here, too, the basal diet provided 0.5 g
of nitrogen from fruits and vegetables. Again, the calculated amino acid
intakes resemble amazingly the requirement values of Rose (Column 9,
Table 6), except for the sulfur-containing amino acids and tryptophan.
These amounts did not, however, maintain nitrogen equilibrium in some
of the subjects.

Values for the sulfur-containing amino acid requirements are not as
satisfactory as those for most others (Hegsted, 1963), and the estimates
derived from subjects fed amino acid diets appear high relative to those
for other amino acids (FAO/WHO, 1965). The calculations summarized
in Table 7 (Columns 11 and 12) support that view; and in a study by
Reynolds et al. (1958), several female subjects were in nitrogen equilib-

TABLE 7 Daily Intakes of Amino Acids Calculated from Protein Intakes in Nitrogen Balance Studies

	Milk[a]	White Flour[a]	Soybean Flour[a]	Whole Egg 8.1g[b]	Whole Egg 10.0g[c]	Rice[d]	Corn[e]	Milk[f] Undiluted	Milk[f] 20% Gly, DAC	Flour + Milk[g]	Minimum Intake for N-Equilibrium or Positive Balance M	F
Total N (g)	3.59	6.19	3.74	6.5	13.0	12.5	12.7	4.3	4.3	8.15		
Protein N (g)	3.59	6.19	3.74	1.3	1.6	5.0	5.0	4.3	3.4	6.02		
Subjects	F	F	F	F	F,M	M	M	M	M	M,F		
Amino acid (mg/day)											*mg/kg*	
Ile	1,460	1,652	1,256	548	676	1,395	1,445	1,707	1,365	1,380	10.5	9.8
Leu	2,246	2,718	1,802	726	897	2,565	4,050	2,616	2,093	2,130	14.0	13.2
Lys	1,780	805	1,477	528	652	1,175	900	2,013	1,610	1,010	10.2	9.6
Met	560	462	314	259	319	535	580	642	514	490	–	–
Cys	205	705	415	193	238	405	405	276	221	620	–	–
TSAA	*765*	*1,167*	*729*	*452*	*557*	*940*	*985*	*918*	*735*	*1,110*	*8.7*	*8.2*
Phe	1,109	1,960	1,155	477	588	1,495	1,420	1,332	1,065	1,490	–	–
Tyr	1,166	1,205	744	355	438	1,360	1,910	1,358	1,086	1,180	–	–
TAAA	*2,275*	*3,165*	*1,899*	*832*	*1,026*	*2,855*	*3,330*	*2,690*	*2,151*	*2,670*	*16.0*	*15.1*
Thr	1,055	1,005	920	411	507	1,165	1,245	1,224	979	930	7.9	7.5
Trp	323	440	322	136	168	320	190	385	307	380	2.6	2.5
Val	1,572	1,550	1,226	612	756	2,080	1,595	1,849	1,479	1,480	11.8	11.1

[a]Bricker et al., 1945.
[b]Swendseid et al., 1960.
[c]Swendseid et al., 1959.
[d]Chen et al., 1967.
[e]Kies et al., 1967.
[f]Scrimshaw et al., 1969.
[g]Goyal and Clark, 1964.

44

rium with methionine intakes of only 300 mg/day. Leverton (1959) concluded that the total sulfur–amino acid requirement was 550 mg/day or about 9.5 mg/kg of body wt/day. There is also a greater discrepancy between the requirements of infants and adults for the sulfur-containing amino acids than for others—a further indication that the adult values are disproportionately high (Harper, 1973). The same appears to be true for tryptophan. Indeed, in the studies of Rose et al. (1954) and Fisher et al. (1963), few individual male subjects required tryptophan in excess of 150 mg/day; and values reported by Fisher et al. (1969) for female subjects were considerably lower.

The other striking feature of Table 7 is that, in trials in which nitrogen equilibrium, or slight positive balance, was attained with low nitrogen intakes (Columns 1 and 3) or with minimal quantities of cereal grain proteins (Columns 2, 6, 7, and 10), amino acid intakes exceeded the requirements listed in Columns 6 and 8 of Table 6. In the experiments by Kies et al. (1967) and Chen et al. (1967) (Table 7, columns 6 and 7), 6 g of nitrogen from rice proteins was required to maintain the subjects in nitrogen equilibrium when total nitrogen intakes were reduced to 6.6 g/day; and 7 g of nitrogen from corn proteins was required when total nitrogen intake was reduced to 7.7 g/day.

The values reported in Columns 8 and 9 (Table 7) are representative of several trials in which marginally adequate amounts of nitrogen from milk, meat, or egg were fed (Huang et al., 1966; Scrimshaw et al., 1966, 1969). In these trials, many subjects were in slight negative nitrogen balance, despite intakes of essential amino acids in excess of the estimated requirements (Columns 7 and 9, Table 6) and in excess of the intakes of subjects fed low amounts of egg protein in high-nitrogen diets (Column 5, Table 7). When Scrimshaw and associates diluted the diets with 20–30 percent of nonspecific nitrogen, the subjects exhibited no further increase in nitrogen loss, suggesting that their total nitrogen intake, rather than the intake of any specific amino acid, was inadequate. Minimal intakes of egg protein that maintain adult men in nitrogen equilibrium (Calloway and Margen, 1971) provide quantities of essential amino acids considerably in excess of the estimated requirements.

Studies by Swendseid et al. (1959, 1960) and by Kies et al. (1967) suggest that with a nitrogen intake on the order of 10 g/day, quantities of essential amino acids approaching the estimated requirements will maintain individual human adults in nitrogen equilibrium. Other investigations by Romo and Linkswiler (1969) and Clark et al. (1967) suggest that, when nitrogen intake is only about 6 g/day, nitrogen retention is improved if intakes of essential amino acids are increased.

Taken altogether, the results of these and other studies of the

TABLE 8 Essential Amino Acid Intakes of Infants with Satisfactory Weight Gain

| | Cow's Milk Formula (Snyderman et al., 1962) | | | | Milk or Formula (Fomon, 1961a) | | | Minimum by Method of Harte and Travers (1947) |
| | A[a] | | B[b] | | Human Milk | Cow's Milk | Soybean Formula | |
	Mean	Range	Mean	Range				
Protein intake[c] (g/kg)	2.35	1.75–3.10	1.19	1.10–1.32	1.50	1.46	1.73	
N intake (g/kg)	0.38	0.28–0.49	0.73	0.68–0.80	0.24	0.23	0.28	
Amino acid (mg/kg)								
His	63	47–82	32	30–35	33	39	42	33
Ile	153	114–199	78	73–85	83	94	94	83
Leu	235	175–307	121	113–131	136	144	135	135
Lys	186	139–243	95	89–104	99	114	111	99
Met	59	44–76	30	28–33	31	36	24	36
Cys	22	16–28	11	10–12	30	13	31	13
TSAA	*81*	*60–104*	*41*	*38–45*	*61*	*49*	*55*	*49*
Phe	116	87–151	60	56–65	65	71	87	65
Tyr	122	91–159	62	58–68	76	75	56	76
TAAA	*238*	*178–310*	*122*	*114–133*	*141*	*146*	*143*	*141*
Thr	110	82–144	57	53–62	68	68	69	68
Trp	34	25–44	17	16–19	25	21	24	21
Val	165	123–215	85	79–92	94	101	92	92

[a] Average of four subjects.

[b] Cow milk protein + glycine and urea.

[c] Protein nitrogen values × 6.25.

amounts of different proteins required to meet nitrogen needs indicate the following: (1) for adult man, and possibly even for children, high-quality proteins provide quantities of essential amino acids in excess of the requirements when they are fed in amounts that meet the nitrogen requirement; (2) intakes of essential amino acids not much higher than the estimated requirements from low-quality cereal grain proteins or from small amounts of high-quality proteins are adequate to maintain nitrogen equilibrium in adults if sufficient total nitrogen is provided; (3) estimates of the amino acid requirements of adults derived from studies on subjects fed amino acid diets are probably low; and (4) the amounts of essential amino acids required in the diet of adult man may be influenced by total nitrogen intake.

Amino Acid Intakes from Proteins by Infants

Calculated intakes of amino acids by infants fed three different sources of protein and growing satisfactorily are given in Table 8. These values were recalculated from the nitrogen intakes, using published values for amino acid composition of foods obtained by chromatographic analyses (Orr and Watt, 1968). The minimum intakes for some amino acids are remarkably similar to values for amino acid requirements determined directly (Table 6, Column 1); but, in general, the values calculated from intakes of protein are lower. It should be pointed out that the infants studied by Fomon (1961) were 4.5–6 months of age, whereas many of those examined in the amino acid requirement studies (Holt and Snyderman, 1965) were younger. Phenylalanine requirement, for example, is known to fall considerably during the first 6 months of life (Holt and Snyderman, 1967).

EVALUATION OF REQUIREMENT STUDIES

Estimates of nitrogen losses in adults indicate that a minimum of 49 mg of N/kg of body weight must be replaced daily, but figures from nitrogen balance studies indicate that adult subjects must consume 75 mg of N/kg of body weight from high-quality proteins to achieve nitrogen equilibrium. These observations imply that efficiency of utilization of nitrogen, even from high-quality proteins, for meeting the maintenance requirement is about 65 percent.

The shortcomings of the nitrogen balance procedure (see p. 31) tend to result in overestimation of nitrogen retention, and the exponential nature of the nitrogen loss curve for subjects fed a protein-free diet probably leads to underestimation of endogenous urinary losses. It is also possible that endogenous losses increase when protein is fed. Despite these difficulties, which may result in underestimation of efficiency of protein utilization in balance studies, the conclusion that nitrogen requirements for maintenance, predicted from summation of nitrogen losses calculated in the accepted way, will be underestimated seems inescapable, as was emphasized by Calloway and Margen (1971). The value of 75 mg of N/kg/day for the average nitrogen requirement of adults, derived from nitrogen balance studies, has proved to be reproducible. Long-term studies, however, of subjects with intakes in the range of the requirements that have been established in shorter studies, accompanied by measurements of body composition, might provide a more reliable value.

Use of the nitrogen balance procedure for estimating requirements of infants and their efficiency of protein utilization poses similar difficulties. Measurement of growth rate in relation to intake of nitrogen provides an alternative approach for estimating requirements of growing subjects that seems more reliable. However, as growth rate slows, growth must be monitored over a longer period of time. As a result, errors in nitrogen intake measurements tend to accumulate.

In measuring efficiency of nitrogen utilization by infants, for ethical and practical reasons the problem of making an appropriate correction for metabolic and endogenous losses cannot be readily solved. It must be recognized that most estimates in this category represent "apparent" utilization rather than "true" utilization and therefore tend to underestimate the true efficiency of nitrogen utilization.

Estimation of amino acid requirements poses all of the problems encountered in attempting to estimate nitrogen requirements, as well as certain others. The estimates for adults that appear most valid (Table 6) are those done by regression analysis (Column 7). However, the estimates are of the amounts required to maintain nitrogen equilibrium rather than to provide a positive balance of about 0.5 g of N/day. Estimates from regression equations, using +0.5-g retention as the criterion (Hegsted, 1963), give much higher values; but they are not consistent with those based on nitrogen equilibrium. It is doubtful, then, with the degree of individual variability observed and with the small number of points at intakes in excess of the amount needed to ensure nitrogen equilibrium, whether an analysis of existing data using this criterion gives reliable values. Nevertheless, since integumental and other minor nitrogen losses were not taken into account in the studies of human amino acid requirements and since no allowance was made for additive errors of the technique, the values listed in Table 6 are, in all likelihood, underestimates.

The estimates for children (Table 6) are from a single series of experiments. In these, the increments between the intakes studied were large (for several amino acids 400 mg/day or more), and the number of subjects was small. The accuracy with which the requirements are defined, therefore, is limited. The intakes selected for study produced distinct positive balances, but it cannot be assumed that lower intakes would not have done so. Whether the values estimated by interpolation are more realistic is open to question, as the requirement pattern still deviates from those for both adults and infants.

When nitrogen intakes of adult men were below the estimated requirement, as in the studies with meat, milk, and egg protein (Table 7, milk), intakes of essential amino acids greatly in excess of the estimated requirements did not permit attainment of nitrogen equilibrium. However, when total nitrogen was increased by providing nonspecific sources of nitrogen and the intake of egg protein was low (Table 7, whole egg), a number of subjects were in positive nitrogen balance, although they had intakes of essential amino acids that did not exceed by much the requirements estimated on the basis of amino acid diets. Clark *et al.* (1963) and Kies *et al.* (1965a) have reported that nitrogen balance improved when amounts of cornmeal that did not support nitrogen equilibrium were supplemented with nonspecific sources of nitrogen. In experiments in which nitrogen intake was held constant, Clark *et al.* (1967) and Romo and Linkswiler (1969) observed that nitrogen retention of young men improved as the quantities of essential amino acids in the diet were increased well above the estimated requirements. Weller *et al.* (1971) have recently reported that intakes of amino acids 30 percent in excess of the requirements estimated by Rose did not maintain N equilibrium in young men consuming 7 g of nitrogen daily. However, Swendseid *et al.* (1959, 1960) achieved nitrogen equilibrium in some subjects that were fed 6.5 or 10 g of nitrogen and given amino acids from egg in amounts near the estimated requirements. Rose and Wixom (1955c) maintained young men in positive nitrogen balance by giving them less than 6 g of nitrogen and intakes of essential amino acids that were double the requirements. These observations also tend to support the view that requirements for amino acids, when estimated using amino acid diets (Table 6), are probably low. They also indicate the need for investigation of the relationship between requirements for essential amino acids and total nitrogen requirement for maintenance.

The observations on intakes of amino acids by infants that are growing well on low-protein formulas (Table 8) support the view that estimates of amino acid requirements obtained from studies of infants fed amino acid diets tend to be high. However, it should be noted that Holt *et al.* (1960), as did Rose (1957) earlier, accepted as the require-

TABLE 9 Comparison of Amino Acid Requirements as Determined Directly or Indirectly (mg/kg/day)

Amino Acid	Adults Requirement (Table 6, Column 7)	Egg Protein (g) 10.0	12.5	15	Corn Protein 31.3 g	Children Requirement (Table 6)	Infants Requirement (Table 6)	Cow's Milk[a] (Table 8, Cols. 1 and 3) Undiluted	Diluted	Minimum (Table 8, Cols. 5-7)
His							34	63	32	33
Ile	9.5	9.4	11.8	14.3	20.6	28	119	153	78	83
Leu	12.5	12.6	15.7	18.8	57.9	42	150	235	121	135
Lys	9.4	9.1	11.4	13.7	12.9	44	103	186	95	99
Met	(3.3–12.1)	4.5	5.6	6.7	8.3	22	45	59	30	36
Cys	(8.6–0.2)	3.4	4.2	5.0	5.8	Absent	80	22	11	13
TSAA	12.1	7.9	9.8	11.7	14.1	22	125	81	41	49
Phe	12.1	8.2	10.3	12.4	20.3	22	90	116	60	65
Tyr	Absent	6.2	7.7	9.2	27.3	Absent	Present	122	62	76
TAAA	12.1	14.4	18.0	21.6	47.6	22	90+	238	122	141
Thr	6.5	7.1	8.9	10.7	17.8	28	87	110	57	68
Trp	2.9	2.3	2.9	3.5	2.7	3.3	22	34	17	21
Val	10.7	10.6	13.2	15.8	22.8	25	105	165	85	92

[a]Cow's milk undiluted; diluted with gly and urea nitrogen.

50

ment the highest value for any one individual and that the infants in their study were somewhat younger than those used in a study (Fomon, 1961a) in which requirements were estimated from intakes of formulas. The last column of Table 8, nevertheless, indicates that infants from 4.5 to 6 months of age can maintain satisfactory growth rates with intakes of several amino acids from milk proteins that are well below the requirements estimated by Holt *et al.* (1960). Again, the individual variability of requirements should be recognized. Fomon and Filer (1967) reported that three infants in the group studied failed to thrive on intakes of nitrogen and amino acids, despite the fact that their intakes overlapped ones that supported satisfactory rates of growth for 19 other subjects.

Table 9 summarizes intakes of amino acids from protein-containing diets that proved adequate for infants and adults, as well as requirements determined in studies on subjects consuming amino acid diets. The close similarity between the amino acid requirements of adults (as determined using amino acid diets) and the amounts provided by 10 g of whole egg proteins (an amount that was demonstrated to maintain several subjects in nitrogen equilibrium when total nitrogen intake was above the estimated requirement) is apparent (Columns 1 and 2). The major discrepancy appears in relation to the sulfur-containing amino acids; the possibility of overestimating the requirement for these substances has been mentioned. Also, the amount of tryptophan supplied by 10 g of egg protein is lower than the estimated adult requirement.

Most of the amino acid intakes from 31.3 g of corn protein (an amount that maintained adult subjects in positive nitrogen balance when nitrogen intake was high) exceeded the estimated requirements. Kies *et al.* (1967) have shown that tryptophan is not limiting for men fed this amount of corn in an otherwise adequate diet that is high in total nitrogen; the requirements of all but one individual studied by Rose *et al.* (1954) were satisfied by 2.1 mg of tryptophan/kg/day. If the digestibility of corn protein is less than that of high-quality protein, the tryptophan value for subjects consuming corn would be somewhat lower. Young *et al.* (1971) concluded that the tryptophan requirement of young men was between 2 and 2.6 mg/kg on the basis of nitrogen balance experiments, although plasma amino acid measurements suggested that it might be somewhat higher. It therefore seems, as has been suggested (FAO/WHO, 1965), that among adults the commonly accepted estimates of requirements for tryptophan and sulfur-containing amino acids are high in relation to those for other amino acids.

The values for infants fed diluted cow's milk (Snyderman *et al.*, 1962)

TABLE 10 Ratios and Patterns of Amino Acid Requirements

Amino Acid	Requirement (mg/kg)			Ratio			Pattern			
	Adult[a]	Child[b]	Infant[c] (4–6 mo)	Infant/ Adult	Infant/ Child	Child/ Adult	Adult	Child ÷ 3	Infant ÷ 9.5	Mean
Protein	470	750	1,500	3.2	2.0	1.6				
His			33							
Ile	9.5	28	83	8.7	3.0	3.0	9.5	9.3	8.7	9.2
Leu	12.5	42	135	10.8	3.2	3.3	12.5	14.0	14.2	13.6
Lys	9.4	44	99	10.5	2.2	4.7	9.4	14.7	10.4	11.5
TSAA	7.9	22	49	6.2	2.2	2.8	7.9	7.3	5.2	6.8
TAAA	12.1	22	141	11.6	6.4	1.8	12.1	7.3	14.8	11.4
Thr	6.5	28	68	10.5	2.4	4.3	6.5	9.3	7.2	7.7
Trp	2.3	3.3	21	9.1	6.4	1.4	2.3	1.1	2.2	1.9
Val	10.7	25	92	8.6	3.7	2.3	10.7	8.3	9.7	9.6
				Mean 9.5						
Total IAA	78	214	721							
IAA as % protein	17	30	48							

[a] From Table 9, Column 1, except TSAA and Trp from Column 2.   [b] From Table 9, Column 6.   [c] From Table 8, Column 8.

and those for infants fed low-protein formulas (Fomon, 1961a) corre-
spond closely (Table 9, Columns 9 and 10). All of these values are below
the requirements estimated on the basis of amino acid diets (Column 7).
Here, too, the major discrepancy has to do with the sulfur-containing
amino acids, but the requirement for methionine was determined only
with diets that provided a large excess of cystine. Quantities of milk
that permitted satisfactory growth provided much smaller amounts of
sulfur-containing amino acids than did the diets used in the requirement
studies. In view of the satisfactory performance of infants when provided these intakes of milk, the average amino acid requirements of
infants 4–6 months of age probably do not exceed the values in Table 9
(Column 10).

REQUIREMENTS, RATIOS, AND PATTERNS

The estimated amino acid requirements of adults, children, and infants
are listed in Table 10 (Column 1–3). Those of the sulfur-containing
amino acids and tryptophan (for adults) are based on intakes from
quantities of egg protein that meet the requirements for the other
amino acids and are lower than Hegsted's (1963) estimates. Certain
evidence suggesting that these estimates are high in relation to those for
other amino acids has already been discussed.

The ratios given in Columns 4–6 provide a comparison of nitrogen
and amino acid requirements for different age groups. By maturity, the
nitrogen (protein) requirement falls to one-third of that at infancy;
most amino acid requirements apparently fall to one-eighth or one-tenth
(Holt *et al.*, 1960; Holt and Snyderman, 1961). The magnitude of the
fall may be somewhat exaggerated if the estimates of amino acid requirements of adults are generally low, a point discussed above. Conversely, the fall between infancy and childhood may be underestimated
if the values for children are high. The infant's requirement for phenylalanine falls by at least 50 percent during the first year of life (Holt and
Snyderman, 1967); and, since the nitrogen requirement for growth
represents but a small part of the total nitrogen requirement of the
10–12-year-old child, the amino acid requirements of children would
generally be expected to approach those for adults, as has been shown
for tryptophan and the aromatic amino acids. The pattern of change
with age in amino acid requirements is obviously not established; the
lack of agreement among the ratios for several amino acids within each
column in the table emphasizes the need for more information on this
subject.

If the requirements for all of the amino acids fell with age at the same rate and if the estimates were highly accurate, the ratios within each column (Table 10, Columns 4–6) would be identical. Again, the major discrepancy between infant and adult values appears to pertain to the requirements for the sulfur-containing amino acids. In view of differences in the roles of different amino acids in the body and the different proportions of various proteins being synthesized at different ages, requirements might well not be expected to fall uniformly with age; nevertheless, it seems unlikely that a discrepancy as large as that seen for the sulfur-containing amino acids would occur. The low infant/adult ratio for these substances suggests that the adult requirement is overestimated as compared to requirements for the other amino acids, even if the lower value based on intakes of egg protein is substituted. Viewed teleologically, egg is the protein source for the developing chick. The sulfur-containing amino acid requirements of chicks are relatively greater than those of mammals (Almquist, 1948). It would therefore not be surprising if the amount of sulfur-containing amino acids in egg was disproportionately high for mammals, especially for mature mammals. They are not the limiting amino acids in egg proteins, even for the growing rat.

Amino acid requirements of infants and adults are based on more detailed information than are those for children; the infant/adult ratios are therefore probably the most reliable of the three sets. Comparisons may well indicate something about discrepancies among amino acid requirements at different ages. The differences in the infant/child and child/adult ratios for the total aromatic amino acids and for tryptophan suggest that the requirements of these for children may be underestimated, especially as there is less discrepancy between the requirements of the adult and the infant. Discrepancies in the ratios for lysine, total sulfur-containing amino acids, and threonine suggest that the requirements of children for these amino acids, especially that for lysine, may be overestimated. These observations further indicate the need for more information about amino acid requirements between infancy and adulthood.

A further feature illustrated by comparison of Columns 1 to 3 is that total essential amino acid requirements of adults are low. As listed, they represent only 17 percent of the total protein requirement. Even if the listed requirements are as much as 30 percent low—because they were estimated to maintain nitrogen equilibrium rather than slight positive nitrogen balance—the higher estimates would represent only about 22 percent of the protein requirement. For infants 4–6 months of age, the essential amino acid requirements based on intakes of human and cow's milk formulas (1.5 g/kg of body wt/day) that supported satisfactory growth represent 48 percent of the total protein requirement. The observations of Snyderman *et al.* (1962), that small amounts of milk protein supplemented with a source of nonspecific nitrogen will support satisfactory growth of infants (Table 5), indicate that some of the amino acid requirements of infants are below those listed. The low essential amino acid requirements for maintenance would account for the effectiveness of small quantities of such protein sources as egg (Swendseid *et al.*, 1959, 1960; Scrimshaw *et al.*, 1966), cow's milk (Scrimshaw *et al.*, 1969), beef (Huang *et al.*, 1966), and cereal proteins (Chen *et al.*, 1967; Kies *et al.*, 1967), supplemented with nonspecific sources of nitrogen for adults. These observations imply that for adult man total nitrogen becomes limiting before the essential amino acids, as intake of high-quality proteins decreases.

The discrepancies mentioned in discussing the ratios for the age groups are once more evident in the patterns given in Columns 7–9 of Table 10. Surprisingly, considering the limitations of the estimates of amino acid requirements, there is much agreement among the patterns for the age groups, especially for the infant and adult, when the infant requirements are divided by 9.5 and those for the children by 3, the average infant/adult and child/adult ratios, respectively. The amino acid requirements of the infant are based on observed intakes of human or cow's milk. Since the sulfur-containing amino acid requirement of infants is based on intakes of cow's milk proteins, which is utilized as efficiently as any protein for adult maintenance (Sumner and Murlin, 1938; Bricker *et al.*, 1945), the discrepancy between the infant and adult values for sulfur-containing amino acids is probably due to overestimation of the adult requirement. There is a distinct need for reinvestigation of adult needs for these amino acids. The pattern for the child diverges in several respects from the other two, especially for lysine, threonine, total aromatic amino acids, and tryptophan.

These comparisons suggest that previous amino acid scoring patterns for evaluating proteins, based on amino acid requirements (FAO, 1957; FNB, 1959) are inadequate, particularly since the adult requirements for sulfur-containing amino acids and tryptophan appear to have been overestimated and since these amino acids are two of the three amino acids most likely to be deficient in human diets. The use of the amino acid composition of egg or human milk proteins as standards for assessing protein quality for man (FAO/WHO, 1965) appears invalid, in view of their high sulfur-containing amino acid and tryptophan content. Nevertheless, despite their inadequacies, past efforts to provide standard reference amino acid patterns have focused attention on nutritional

practices, have established the value of amino acid scoring as a procedure for obtaining an approximation of the nutritional quality of proteins, and have stimulated additional research.

Various amino acid scoring patterns that have been proposed as
standards for evaluating the nutritional quality of food proteins by the
Chemical Score procedure are shown in Table 11, together with the
pattern of amino acid requirements of infants expressed as mg/g of protein, derived from information on intakes of human or cow's milk that
were adequate for growth (see Table 8). In the last column, a new amino
acid scoring pattern is proposed. In devising this, it has been assumed
that, since amino acid requirements evidently fall more rapidly with age
than does the requirement for total nitrogen and since cow's milk protein can evidently be diluted about 30 percent with nonspecific sources
of nitrogen without loss of nutritional value (Table 5), food proteins
that contain the amounts of amino acids listed in Column 5 in 1.3 g of
total protein will be of high nutritional value. In the scoring pattern
given in Column 6, this relationship is expressed as amount of amino
acid required per gram of protein. Although the proposed pattern differs considerably from the 1957 FAO pattern, the most critical features
are the lower values for the sulfur-containing amino acids and tryptophan. If, as the information reviewed indicates, 1.5 g of protein containing the proposed amounts of these two amino acids will meet the
needs of the infant 4–6 months of age (Table 10), then such a protein
fed in an amount that meets the nitrogen requirement should be more
than adequate for older age groups. It would provide all amino acids
except the sulfur-containing ones in great excess of the estimated adult

TABLE 11   Amino Acid Patterns Proposed for Evaluating the Nutritional Quality
of Proteins (mg/g)

Amino Acid	FAO 1957	FAO 1965 Human Milk[a]	Egg[a]	Cow's Milk[a]	Minimum (Cols. 2–4)	Proposed Pattern
His		22	24	27	22	17
Ile	42	55	66	65	55	42
Leu	48	91	91	100	91	70
Lys	42	66	66	80	66	51
TSAA	42	41	55	34	34	26
TAAA	56	95	101	100	95	73
Thr	28	45	50	47	45	35
Trp	14	16	18	14	14	11
Val	42	62	74	70	62	48
Total	314	493	545	537	484	373

[a]Orr and Watt, 1968.

requirements. The proposed pattern has not been tested directly; it should be considered as tentative and be subjected to critical evaluation.

AMINO ACID REQUIREMENTS AND PROTEIN QUALITY

If the requirements of adult man for essential amino acids represent as small a proportion of the nitrogen requirement as that shown in Table 10, many proteins of low nutritional quality should be as adequate as high high-quality proteins. The differences observed are not as great as would be anticipated from measurements of Net Protein Utilization in laboratory animals, since soybean flour is apparently as adequate as milk for adults (Table 2). Differences are observed, nevertheless, e.g., the requirement for white flour proteins is 50 percent greater than that for milk proteins. Assuming 4.9 g/day as the nitrogen requirement of a 70-kg man, that amount of nitrogen from white flour would carry with it about 450 mg of lysine, the limiting amino acid, assuming a high degree of refinement (i.e., extraction rate) of the flour. The lysine requirement of a 70-kg man, estimated experimentally, would be predicted at about 660 mg/day. Thus the lysine content of an amount of flour that just met the nitrogen requirement would be inadequate, and a response to increased flour intake would be anticipated. The amount of nitrogen from white flour required to support nitrogen equilibrium in a 70-kg man has been estimated at about 7.4 g/day, an amount that would provide about 680 mg of lysine. The higher requirement for nitrogen from flour than from egg or milk is thus well predicted from a knowledge of amino acid requirements. If the amino acid scoring pattern had been used instead of the actual lysine requirement for assessing the adequacy of white flour for adults, need for a much greater nitrogen intake would have been predicted. Thus, although the scoring pattern may be appropriate for assessing protein quality for young children, it will underestimate protein quality for adults.

The possibility should be considered, as knowledge of the amino acid composition of foods and diets increases and as accuracy of human amino acid requirements at different ages improves, of using direct comparisons of the diet content to the requirement, expressed per gram of protein or nitrogen, as the method of choice in assessing protein quality.

REFERENCES

Almquist, H. J. 1948. The amino acid requirements of avian species. Pages 221–235 in M. Sahyun, ed. Proteins and amino acids in nutrition. Reinhold, New York.

Anderson, H. L., M. B. Heindel, and H. Linkswiler. 1969. Effect of nitrogen balance of adult man of varying source of nitrogen and level of caloric intake. J. Nutr. 99:82–90.

Anonymous. 1964. Histidine requirement in infancy. Nutr. Rev. 22:114–115.

Ashworth, U. S. 1935. Endogenous nitrogen and basal energy relationships during growth. Mo. Agric. Exp. Stn. Res. Bull. 223:3–20.

Ashworth, U. S., and G. R. Cowgill. 1938. Body composition as a factor governing the basal heat production and the endogenous nitrogen excretion. J. Nutr. 15:73–81.

Beach, E. F., S. S. Bernstein, and I. G. Macy. 1941. Intake of amino acids by breast-milk-fed infants and amino acid composition of cow's milk and human milk. J. Pediatr. 19:190–200.

Benevenga, N. J., A. E. Harper, and Q. R. Rogers. 1968. Effect of an amino acid imbalance on the metabolism of the limiting amino acid in the rat. J. Nutr. 95:434–444.

Bricker, M. L., and J. M. Smith. 1951. A study of the endogenous nitrogen output of college women, with particular reference to use of the creatinine output in the calculation of the biological values of the protein of egg and of sunflower seed flour. J. Nutr. 44:553–573.

Bricker, M., H. H. Mitchell, and G. M. Kinsman. 1945. The protein requirements of adult human subjects in terms of the protein contained in individual foods and food combinations. J. Nutr. 30:269–283.

Bricker, M. L., R. F. Shively, J. M. Smith, H. H. Mitchell, and T. S. Hamilton. 1949. The protein requirements of college women on high cereal diets with observations on the adequacy of short balance periods. J. Nutr. 37:163–183.

Burrill, L. M., and C. Schuck. 1964. Phenylalanine requirements with different levels of tyrosine. J. Nutr. 83:202–208.

Calloway, D. H., and S. Margen. 1971. Variation in endogenous nitrogen excretion and dietary nitrogen utilization as determinants of human protein requirement. J. Nutr. 101:205–216.

Calloway, D. H., A. C. F. Odell, and S. Margen. 1971. Sweat and miscellaneous nitrogen losses in human balance studies. J. Nutr. 101:775–786.

Chan, H., and J. C. Waterlow. 1966. The protein requirement of infants at the age of about one year. Br. J. Nutr. 20:775–782.

Chen, S. C., H. M. Fox, and C. Kies. 1967. Nitrogenous factors affecting the adequacy of rice to meet the protein requirements of human adults. J. Nutr. 92:429–434.

Clark, H. E., M. A. Kenney, A. F. Goodwin, K. Goyal, and E. T. Mertz. 1963. Effect of certain factors on nitrogen retention and lysine requirements of adult human subjects. IV. Total nitrogen intake. J. Nutr. 81:223–229.

Clark, H. E., K. Fugate, and P. E. Allen. 1967. Effect of four multiples of a basic mixture of essential amino acids on nitrogen retention of adult human subjects. Am. J. Clin. Nutr. 20:233–242.

Cuthbertson, D. P., and W. S. W. Guthrie. 1934. The effect of variations in protein and salt intake on the nitrogen and chloride content of sweat. Biochem. J. 28:1444–1453.

Darke, S. J. 1960. The cutaneous loss of nitrogen compounds in African adults. Br. J. Nutr. 14:115–119.

DeMaeyer, E. M., and H. L. Vanderborght. 1961. Determination of the nutritive value of different protein foods in the feeding of African children. Pages 143–155 *in* Progress in meeting protein needs of infants and preschool children. Publ. 843. National Academy of Sciences, Washington, D.C.

Deuel, H. J., Jr., I. Sandiford, K. Sandiford, and W. M. Boothby. 1928. A study of the nitrogen minimum. The effect of 63 days of a protein-free diet on the nitrogen partition products in the urine and on heat production. J. Biol. Chem. 76: 391–406.

FAO (Food and Agriculture Organization). 1957. Protein requirements. FAO Nutr. Stud. No. 16. FAO, Rome. 52 pp.

FAO (Food and Agriculture Organization). 1970. Amino-acid content of foods and biological data on proteins, FAO Nutr. Stud. No. 24. FAO, Rome. 285 pp.

FAO/WHO (Food and Agriculture Organization/World Health Organization). 1965. Protein requirements. WHO Tech. Rep. Ser. No. 301; FAO Nutr. Meet. Rep. Ser. No. 37. WHO, Geneva. 71 pp.

Fisher, H., M. K. Brush, R. Shapiro, J. P. H. Wessels, C. G. Berdanier, P. Grimminger, and E. R. Sostman. 1963. Amino acid balance in the adult: High nitrogen–low tryptophan diets. J. Nutr. 81:230–234.

Fisher, H., M. K. Brush, and P. Grimminger. 1969. Reassessment of amino acid requirements of young women on low nitrogen diets. I. Lysine and tryptophan. Am. J. Clin. Nutr. 22:1190–1196.

FNB (Food and Nutrition Board, National Research Council). 1959. Evaluation of protein nutrition. Publ. 711. National Academy of Sciences, Washington, D.C. 61 pp.

FNB (Food and Nutrition Board, National Research Council). 1968. Recommended dietary allowances, 7th ed. National Academy of Sciences, Washington, D.C. 101 pp.

Fomon, S. J. 1959. Comparative study of human milk and a soya bean formula in promoting growth and nitrogen retention by infants. Pediatrics 24:577–584.

Fomon, S. J. 1960. Comparative study of adequacy of protein from human milk and cow's milk in promoting nitrogen retention by normal full-term infants. Pediatrics 26:51–61.

Fomon, S. J. 1961a. Factors influencing retention of nitrogen by normal full-term infants. Pages 343–353 *in* Progress in meeting protein needs of infants and preschool children. Publ. 843. National Academy of Sciences, Washington, D.C.

Fomon, S. J. 1961b. Nitrogen balance studies with normal full-term infants receiving high intakes of protein. Pediatrics 28:347–361.

Fomon, S. J. 1967. Infant nutrition. W. B. Saunders, Philadelphia. 299 pp.

Fomon, S. J., and L. J. Filer, Jr. 1967. Amino acid requirements for normal growth. Pages 391–402 *in* W. L. Nyhan, ed. Amino acid metabolism and genetic variation. McGraw–Hill, New York.

Fomon, S. J., and C. D. May. 1958. Metabolic studies of normal full-term infants fed pasteurized human milk. Pediatrics 22:101–114.

Fomon, S. J., E. M. DeMaeyer, and G. M. Owen. 1965. Urinary fecal excretion of endogenous nitrogen by infants and children. J. Nutr. 85:235–246.

Fomon, S. J., L. J. Filer, Jr., L. N. Thomas, Q. R. Rogers, and A. M. Proksch. 1969. Relationship between formula concentration and rate of growth of normal infants. J. Nutr. 98:241–254.

Freyberg, R. H., and R. L. Grant. 1937. Loss of minerals through the skin of normal humans when sweating is avoided. J. Clin. Invest. 16:729–731.

Gopalan, C., and B. S. Narasinga Rao. 1966. Effect of protein depletion on urinary nitrogen excretion in undernourished subjects. J. Nutr. 90:213–218.

Goyal, N. O., and H. E. Clark. 1964. Nitrogen retention of adult subjects on a diet containing whole wheat flour and nonfat dry milk. J. Nutr. Diet. 1:288–292.

Grindley, H. S., and H. H. Mitchell. 1917. Studies in nutrition. An investigation of the influence of saltpeter on the nutrition and health of man with reference to its occurrence in cured meats. Vol. 1: Discussion and interpretation of the biochemical data. University of Illinois, Urbana. 544 pp.

Harper, A. E. 1965. Effect of variations in protein intake on enzymes of amino acid metabolism. Can. J. Biochem. 43:1589–1603.

Harper, A. E. 1973. Amino acid requirements and plasma amino acids. *in* H. Brown, ed. Protein Nutrition. C. C. Thomas, Springfield, Ill.

Harte, R. A., and J. J. Travers. 1947. Human amino acid requirements. Science 105:15–16.

Hawley, E. E., J. R. Murlin, E. S. Nasset, and T. A. Szymanski. 1948. Biological values of six partially-purified proteins. J. Nutr. 36:153–169.

Hegsted, D. M. 1957. Theoretical estimates of the protein requirements of children. J. Am. Diet. Assoc. 33:225–232.

Hegsted, D. M. 1963. Variation in requirements of nutrients—amino acids. Fed. Proc. 22:1424–1430.

Hegsted, D. M. 1964. Protein requirements. Pages 135–171 *in* H. N. Munro and J. B. Allison, eds. Mammalian protein metabolism. Vol. 2. Academic Press, New York.

Hegsted, D. M., A. G. Tsongas, D. B. Abbott, and F. J. Stare. 1946. Protein requirements of adults. J. Lab. Clin. Med. 31:261–284.

Holmes, E. G. 1965. An appraisal of the evidence upon which recently recommended protein allowances have been based. World Rev. Nutr. Diet. 5:237–274.

Holt, L. E., Jr., and S. E. Snyderman. 1961. The amino acid requirements of infants. J. Am. Med. Assoc. 175:100–103.

Holt, L. E., Jr., and S. E. Snyderman. 1965. Protein and amino acid requirements of infants and children. Nutr. Abstr. Rev. 35:1–13.

Holt, L. E., and S. E. Snyderman. 1967. The amino acid requirements of children. Pages 381–390 *in* W. L. Nyhan, ed. Amino acid metabolism and genetic variation. McGraw–Hill, New York.

Holt, L. E., Jr., P. Gyorgy, E. L. Pratt, S. E. Snyderman, and W. M. Wallace. 1960. Protein and amino acid requirements in early life. New York University Press, New York. 63 pp.

Holt, L. E., Jr., E. Halac, Jr., and C. N. Kajdi. 1962. The concept of protein stores and its implications in diet. J. Am. Med. Assoc. 181:699–705.

Huang, P. C., V. R. Young, B. Cholakos, and N. S. Scrimshaw. 1966. Determination of the minimum dietary essential amino acid-to-total nitrogen ratio for beef protein fed to young men. J. Nutr. 90:416–422.

Kies, C., E. Williams, and H. M. Fox. 1965a. Determination of first limiting nitrogenous factor in corn protein for nitrogen retention in human adults. J. Nutr. 86:350–356.

Kies, C., E. Williams, and H. M. Fox. 1965b. Effect of "non-specific" nitrogen intake on adequacy of cereal proteins for nitrogen retention in human adults. J. Nutr. 86:357–361.

Kies, C., E. Williams, and H. M. Fox. 1967. Effect of nonspecific nitrogen supplementation on minimum corn protein requirement and first-limiting amino acid for adult men. J. Nutr. 92:377–383.

Leverton, R. M. 1959. Amino acid requirements of young adults. Pages 477–506 *in* A. A. Albanese, ed. Protein and amino acid nutrition. Academic Press, New York.

Lusk, G. 1920. The science of nutrition, 3rd ed. W. B. Saunders, Philadelphia. 450 pp.

Martin, C. J., and R. Robinson. 1922. The minimum nitrogen expenditure of man and the biological value of proteins for human nutrition. Biochem. J. 16:407–447.

Miller, D. S., and P. R. Payne. 1969. Assessment of protein requirements by nitrogen balance. Proc. Nutr. Soc. 28:225–233.

Mitchell, H. H. 1924. The biological value of proteins at different levels of intake. J. Biol. Chem. 58:905–922.

Mitchell, H. H. 1948. The biological utilization of proteins and protein requirements. Pages 46–81 in M. Sayhun, ed. Proteins and amino acids in nutrition. Reinhold, New York.

Mitchell, H. H. 1964. Comparative nutrition of man and domestic animals. Vol. 2. Academic Press, New York. 840 pp.

Mitchell, H. H., and M. Edman. 1962. Nutritional significance of the dermal losses of nutrients in man, particularly of nitrogen and minerals. Am. J. Clin. Nutr. 10:163–172.

Mitchell, H. H., and T. S. Hamilton. 1949. The dermal excretion under controlled environmental conditions of nitrogen and minerals in human subjects, with particular reference to calcium and iron. J. Biol. Chem. 178:345–361.

Mueller, A. J., and W. M. Cox, Jr. 1947. Comparative nutritive value of casein and lactalbumin for man. J. Nutr. 34:285–294.

Munro, H. N. 1964. General aspects of the regulation of protein metabolism by diet and by hormones. Pages 381–481 in H. N. Munro and J. B. Allison, eds. Mammalian protein metabolism. Vol. 1. Academic Press, New York.

Murlin, J. R., L. E. Edwards, E. E. Hawley, and L. E. Clark. 1946. Biological value of proteins in relation to the essential amino acids which they contain. I. The endogenous nitrogen of man. J. Nutr. 31:533–554.

Nakagawa, I., T. Takahashi, and T. Suzuki. 1960. Amino acid requirements of children. J. Nutr. 71:176–181.

Nakagawa, I., T. Takahashi, and T. Suzuki. 1961a. Amino acid requirements of children: Isoleucine and leucine. J. Nutr. 73:186–190.

Nakagawa, I., T. Takahashi, and T. Suzuki. 1961b. Amino acid requirements of children: Minimal needs of lysine and methionine based on nitrogen balance method. J. Nutr. 74:401–407.

Nakagawa, I., T. Takahashi, T. Suzuki, and K. Kobayashi. 1962. Amino acid requirements of children: Minimal needs of threonine, valine and phenylalanine based on nitrogen balance method. J. Nutr. 77:61–68.

Nakagawa, I., T. Takahashi, T. Suzuki, and K. Kobayashi. 1963. Amino acid requirements of children: Minimal needs of tryptophan, arginine and histidine based on nitrogen balance method. J. Nutr. 80:305–310.

Nakagawa, I., T. Takahashi, T. Suzuki, and K. Kobayashi. 1964. Amino acid requirements of children: Nitrogen balance at the minimal level of essential amino acids. J. Nutr. 83:115–118.

Nasset, E. S. 1956. Essential amino acids and nitrogen balance. Pages 3–21 in W. H. Cole, ed. Some aspects of amino acid supplementation. Rutgers University Press, New Brunswick, N.J.

Nasset, E. S., and V. H. Gatewood. 1954. Nitrogen balance and hemoglobin of adult rats fed amino acid diets low in L- and D-histidine. J. Nutr. 53:163–176.

Orr, M. L., and B. K. Watt. Reprinted 1968. Amino acid content of foods. Home

Econ. Res. Rep. No. 4 ARS-USDA. U.S. Government Printing Office, Washington, D.C. 82 pp.

Reynolds, M. S., D. L. Steele, E. M. Jones, and C. A. Baumann. 1958. Nitrogen balances of women maintained on various levels of methionine and cystine. J. Nutr. 64:99–111.

Romo, G. S., and H. Linkswiler. 1969. Effect of level and pattern of essential amino acids on nitrogen retention of adult man. J. Nutr. 97:147–153.

Rose, W. C. 1957. The amino acid requirements of adult man. Nutr. Abstr. Rev. 27:631–647.

Rose, W. C., and R. L. Wixom. 1955a. The amino acid requirements of man. XIII. The sparing effect of cystine on the methionine requirement. J. Biol. Chem. 216:763–773.

Rose, W. C., and R. L. Wixom. 1955b. The amino acid requirements of man. XIV. The sparing effect of tyrosine on the phenylalanine requirement. J. Biol. Chem. 217:95–101.

Rose, W. C., and R. L. Wixom. 1955c. The amino acid requirements of man. XVI. The role of the nitrogen intake. J. Biol. Chem. 217:997–1004.

Rose, W. C., W. J. Haines, and D. T. Warner. 1954. The amino acid requirements of man. V. The role of lysine, arginine, and tryptophan. J. Biol. Chem. 206:421–430.

Scrimshaw, N. S., V. R. Young, R. Schwartz, M. L. Piche, and J. B. Das. 1966. Minimum dietary essential amino acid-to-total nitrogen ratio for whole egg protein fed to young men. J. Nutr. 89:9–18.

Scrimshaw, N. S., V. R. Young, P. C. Huang, O. Thanangkul, and B. V. Cholakos. 1969. Partial dietary replacement of milk protein by nonspecific nitrogen in young men. J. Nutr. 98:9–17.

Sherman, H. C., L. H. Gillett, and E. Osterberg. 1920. Protein requirement of maintenance in man and nutritive efficiency of bread protein. J. Biol. Chem. 41:97–109.

Sirbu, E. R., S. Margen, and D. H. Calloway. 1967. Effect of reduced protein intake on nitrogen loss from the human integument. Am. J. Clin. Nutr. 20:1158–1165.

Smith, M. 1926. The minimum endogenous nitrogen metabolism. J. Biol. Chem. 68:15–31.

Smuts, D. B. 1935. The relation between the basal metabolism and the endogenous nitrogen metabolism, with particular reference to the estimation of the maintenance requirement of protein. J. Nutr. 9:403–433.

Snyderman, S. E., L. E. Holt, Jr., J. Dancis, E. Roitman, A. Boyer, and M. E. Balis. 1962. "Unessential" nitrogen: A limiting factor for human growth. J. Nutr. 78:57–72.

Snyderman, S. E., A. Boyer, E. Roitman, L. E. Holt, Jr., and P. H. Prose. 1963. The histidine requirement of the infant. Pediatrics 31:786–801.

Soliman, A-G., and K. W. King. 1969. Metabolic derangements in response of rats to ingestion of imbalanced amino acid mixtures. J. Nutr. 98:255–270.

Sumner, E. E., and J. R. Murlin. 1938. The biological value of milk and egg protein in human subjects. J. Nutr. 16:141–152.

Swendseid, M. E., R. J. Feeley, C. L. Harris, and S. G. Tuttle. 1959. Egg protein as a source of the essential amino acids. Requirement for nitrogen balance in young adults studied at two levels of nitrogen intake. J. Nutr. 68:203–211.

Swendseid, M. E., C. L. Harris, and S. G. Tuttle. 1960. The effect of sources of nonessential nitrogen on nitrogen balance in young adults. J. Nutr. 71:105–108.

Voit, E. 1930. Ueber die Grösse der Erneuerung der Horngebilde beim Menschen; die Nägel. Z. Biol. 90:525–548.

Wallace, W. M. 1959. Nitrogen content of the body and its relation to retention and loss of nitrogen. Fed. Proc. 18:1125–1130.

Waterlow, J. C. 1968. Observations of the mechanism of adaptation to low protein intakes. Lancet 2:1091–1097.

Waterlow, J. C. 1970. Human protein requirements and malnutrition. Pages 15–21 *in* A. E. Bender, R. Kihlberg, B. Löfqvist, and L. Munck, eds. Evaluation of novel protein products. Proc. IBP Werner–Gren Symp., Stockholm, 1968. Pergamon Press, New York.

Waterlow, J. C., and J. M. Stephen, eds. 1957. Human protein requirements and their fulfillment in practice. FAO/WHO/Josiah Macy Jr. Foundation, New York. 193 pp.

Waterlow, J. C., and J. M. Stephen. 1969. Enzymes and the assessment of protein nutrition. Proc. Nutr. Soc. 28:234–242.

Weller, L. A., D. H. Calloway, and S. Margen. 1971. Nitrogen balance of men fed amino acid mixtures based on Rose's requirements, egg white protein and serum free amino acid patterns. J. Nutr. 101:1499–1507.

Woodruff, C. W. 1961. Protein requirements of full-term infants. J. Am. Med. Assoc. 175:100–118.

Yoshida, A., P. M-B. Leung, Q. R. Rogers, and A. E. Harper. 1966. Effect of amino acid imbalance on the fate of the limiting amino acid. J. Nutr. 89:80–90.

Young, V. R., and N. S. Scrimshaw. 1968. Endogenous nitrogen metabolism and plasma free amino acids in young adults given a "protein-free" diet. Br. J. Nutr. 22:9–20.

Young, V. R., M. A. Hussein, E. Murray, and N. S. Scrimshaw, 1971. Plasma tryptophan response curve and its relation to tryptophan requirement in young men. J. Nutr. 101:45–60.

D. M. HEGSTED

Assessment of Protein Quality

The fact that proteins differ greatly in nutritive value was first demonstrated grossly by comparing the performance of animals fed diets containing approximately equal amounts of protein. It is also clear that one can usually compensate for low-quality protein by feeding larger quantities of it. That is, the performance of animals fed a limited quantity of a protein of poor quality will be improved by feeding more of the same protein. The nitrogen balance of human subjects reacts similarly. Finally, it has been established that the nutritional quality of proteins is usually related to amino acid content, since the addition of appropriate amino acids will improve the nutritive value of diets based on poor-quality proteins.

An ideal assay of protein quality would be one such that:

$$g \text{ of protein } A \times f_1 = g \text{ of protein } B \times f_2 = g \text{ of protein } C \times f_3, \quad (1)$$

where f_1, f_2, f_3, etc., are the measures of protein quality. Thus, if one knew the requirement of a given individual for a given protein and the nutritive value of that protein, one could calculate the requirement for another protein, provided its nutritive value were known. It has been commonly assumed that two measures of protein quality, Biological Value (BV) and Net Protein Utilization (NPU), would allow this kind of

64

calculation. Considerable information from animal studies has shown that egg protein is for all practical purposes 100 percent efficient in meeting the protein needs for maintenance of nitrogen balance or for growth, i.e., that the BV and NPU of this protein is 100 percent. Thus, the general approach has been to define the requirement of subjects of differing size, age, and sex in terms of egg protein, and from this value to calculate the requirement for other proteins with known but lower nutritive value (FAO, 1957; FNB, 1959; FAO/WHO, 1965).

It is important to recognize that if the requirement for egg protein [BV = 100 percent or f in Eq. (1) = 1] is X, then the requirement for proteins with a BV of 50 percent, 25 percent, and 12.5 percent would be $2X$, $4X$, and $8X$, respectively. Substantial differences in protein quality at the upper end of the scale have only modest effects upon protein requirements. The uncertainties in estimating requirements may be great enough to overshadow these differences. On the other hand, relatively minor differences in quality have major effects upon the estimated requirement when poor-quality proteins are fed. It is especially important to have reasonably accurate measures of quality for proteins of poor quality, not only because of the effect on requirement of a small difference in quality, but also because protein deficiency is likely to be a practical problem when diets include primarily proteins of poor quality.

It is now evident that the traditional methods of estimating protein quality do not adequately fulfill the expectations expressed in Eq. (1) and are likely to yield increasingly less reliable estimates as the quality falls. They usually overestimate the nutritive value of poor-quality proteins, which leads to an underestimate of the requirement for them. Underestimation of the true protein need is a more serious error than overestimation, since it will result in recommendations that are below requirements. Overestimations are simply wasteful.

BIOLOGICAL VALUE (BV)

Biological Value, defined as the "percentage of absorbed nitrogen retained in the body," has long been considered the method of choice for estimating the nutritive value of proteins (Thomas, 1909; Mitchell, 1924b; and Mitchell and Carman, 1926). A complete evaluation of the dietary protein includes measurements of the Biological Value and the digestibility. These are obtained by measuring the fecal and urinary nitrogen of a subject fed a test protein diet and then correcting for the amounts excreted when a nitrogen-free diet is fed. True digestibility is defined as the percentage of food nitrogen absorbed from the gut:

$$\text{Digestibility} = \frac{I - (F - F_o)}{I} \times 100,$$ (2)

and Biological Value as:

$$\text{BV} = \frac{I - (F - F_o) - (U - U_o)}{I - (F - F_o)} \times 100,$$ (3)

where I is the nitrogen intake of test protein; F, fecal nitrogen; F_o, fecal nitrogen when a nitrogen-free diet is fed (metabolic N); U, urinary nitrogen; and U_o, urinary nitrogen when a nitrogen-free diet is fed (endogenous N).

In practice, Mitchell and Carman (1926) found that the amount of urinary N, when a small amount of high-quality protein was fed, was similar to the endogenous N; and thus they preferred to feed limited amounts of egg protein rather than a nitrogen-free diet, in order to prevent severe weight loss. The basic assumption made in the measurement of Biological Value is that the endogenous N and metabolic N are constant values and can be legitimately subtracted from the test values as shown in Eq. (2) and (3). Limited information suggests that this may not always be true. For example, the excretion of urinary nitrogen in rats and dogs on nitrogen-free diets may be lowered substantially by the administration of methionine (Brush et al., 1947; Johnson et al., 1947), yielding a Biological Value of the latter alone above 100 percent. This has not yet been thoroughly studied and may not happen in man (Cox et al., 1947). Also, Mitchell et al. (1927) found the BV of gelatin to be 20 percent, i.e., 20 percent as satisfactory as the best-quality proteins. Because animals cannot survive on gelatin alone, this must be an overestimate of the real nutritive value.

The overall nutritive value of a protein should be obtained from the Mitchell method as Biological Value × digestibility; this should be identical with NPU as defined below.

NET PROTEIN UTILIZATION (NPU)

Like Biological Value, determination of NPU provides an estimate of nitrogen retention, but does so by measuring the difference between the body nitrogen content of animals fed no protein and those fed a test protein. This value, divided by the amount of protein consumed, is the

NPU, which is defined as the "percentage of the dietary protein retained." Miller (1963) proposed that replicate groups of four weanling
rats housed in group cages be fed either the "protein-free" or the "test"
diet for 10 days. No special merit of these empirically chosen conditions
has been demonstrated. Because a high correlation between body nitrogen and body water content exists in young animals (Miller and Bender,
1955; Dreyer, 1957, 1962; Henry and Toothill, 1962), the substitution
of body water measurements for body nitrogen measurements has been
widely used. Indeed, because certain sampling errors are eliminated,
measurement of body water may be more accurate than measurement
of body nitrogen, as well as being much more convenient and less expen-
sive. If the body composition of young rats is essentially constant during
the few weeks after weaning, then weight gain would be an even more
economical and more accurate measure of protein deposition. Considerable evidence indicates that this is true.

Since both NPU and BV are based upon estimates of "retained nitrogen," they should measure the same thing, except that the calculation
of BV is based on the amount of N absorbed rather than consumed.
Biological Value would thus be expected to be somewhat higher than
NPU. In weanling rats, it is possible, though unproven, that total carcass
analysis is a more accurate measure of "retained nitrogen" than are nitrogen balance determinations. The former is certainly less tedious.
Nitrogen balance measurements must be used in studies on large animals
and man.

AMINO ACID SCORE

Block and Mitchell (1946–47) originally proposed that, because all
amino acids must be present at the site of protein synthesis in adequate
amounts if protein synthesis is to proceed, a deficit of any one of them
would limit synthesis to the same degree as would a comparable deficit
of another. Thus, they suggested that if the composition of an "ideal
protein" were known, i.e., a protein that contained every indispensable
amino acid in sufficient amounts to meet requirements without any excesses, then it should be possible to compute the nutritive value of a
protein by calculating the deficit of each indispensable amino acid in
the test protein as compared to the amount in the "ideal protein." The
"most limiting amino acid," the one in greatest deficit, would presumably determine the nutritive value.

In practice, they suggested the protein in whole egg as the "ideal,"
since this was known to have a BV closely approaching 100. They recog-

nized that it might contain some amino acids in excess of requirements. If so, deficits of these in other proteins calculated by the same procedure would be misleadingly large. That is, the calculated nutritive value would be lower than it actually was. However, Block and Mitchell (1946–47) compared BV's that were thought to have been accurately estimated with "amino acid deficits," calculated using egg protein as the standard, and found a rather high correlation (r = 0.86), which suggested that their procedure was valid (Figure 1).

Amino Acid Score has been widely used since that time. Generally, it has been calculated as the "percentage of adequacy" rather than as the percent deficit. The Food and Agriculture Organization (FAO, 1957), recognizing again that egg proteins might contain various essential amino acids in excess of the amounts required, proposed that Amino Acid Score be calculated from a pattern based upon estimates of amino acid requirements in man. A similar approach was recommended by the

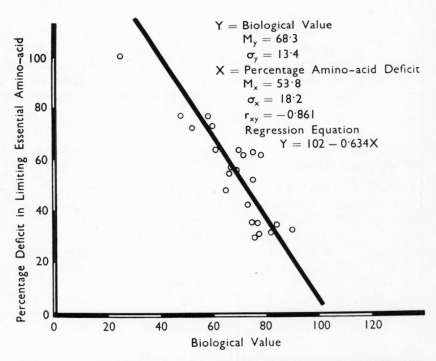

Y = Biological Value
M_y = 68·3
σ_y = 13·4
X = Percentage Amino-acid Deficit
M_x = 53·8
σ_x = 18·2
r_{xy} = −0·861
Regression Equation
$Y = 102 − 0·634X$

FIGURE 1   Original figure from Block and Mitchell (1946–47) relating "Amino Acid Deficit" to Biological Value. Note that 100 percent deficit would apparently yield a BV of about 30 percent.

Food and Nutrition Board (FNB, 1959). However, the second Joint Expert Group of FAO/WHO (1965) concluded that the previously suggested pattern was not appropriate in certain respects and that there was not sufficient information to state that egg, cow's milk, or human milk proteins differed in nutritional quality. They therefore suggested that any of these three patterns might be considered "ideal" for the calculation of Amino Acid Score. Confusion arose, however, because the three proteins differ substantially in amino acid composition. The FAO/WHO also suggested that the ratio of essential amino acid nitrogen to total nitrogen (E:T) was related to, and might be a determinant of, protein quality. Since no method was proposed for combining this ratio with the Amino Acid Score, still further confusion ensued.

CRITIQUE

The use of estimates of protein quality as a base for calculating the amount of protein needed to meet requirements when different diets are consumed requires that the estimate of quality vary in some known fashion, preferably in linear fashion, from 0 to 100 percent utilization. Actually, when Block and Mitchell (1946–47) first proposed the use of Amino Acid Score (Figure 1), they found that BV did not follow the predicted relationship. Rather, the regression line relating BV and Amino Acid Score indicated that proteins completely lacking an essential amino acid, and which therefore would have an Amino Acid Score of zero, seemed to yield a BV of approximately 25 percent. If so, the requirement could be met with such proteins if they were fed at a level providing four times the estimated minimal protein requirement. This simply cannot be true, since it implies that the protein needs could be met without there being a supply of all of the essential amino acids.

This apparent discrepancy between theoretical predictions and experimental results has been largely ignored. Indeed, the FAO committee simply assumed that the relationship must fit theoretical expectations (Figure 2). Obviously, with the scatter of BV data available and uncertainties regarding the amino acid composition of the proteins actually tested, the true relationship was difficult to ascertain. However, it now seems clear that the relationship found by Block and Mitchell (Figure 1) is, in fact, substantially correct. The values presented in Table 1 are plotted in Figure 3 to show the relationship between BV and Amino Acid score. The regression line presented there indicates that a protein of zero score would be predicted to have a BV of 40 percent. If BV were accepted as the true measure of protein quality, proteins of zero

TABLE 1 Quality of Proteins by Various Measures[a]

Protein Source	BV	NPU	PER	Chemical Score
Buckwheat	77			51
Maize	59	51	1.12	41
Oats	65	66	2.19	57
Rice, polished	64	57	2.18	56
Rice, whole	73			57
Sorghum	73		1.78	31
Wheat, whole	65	40	1.53	43
Wheat, germ	74	67	2.53	54
Wheat, gluten	58	39		26
Wheat flour	52		0.60	28
White bread		37	0.89	
Potato	67			34
Beans (various)	58	40	1.48	34
Black gram	70		2.12	41
Lima beans	66	52	1.53	28
Broadbeans	55	48		40
Chickpea	68		1.71	41
Cowpea	57	45		55
Groundnuts	55	43	1.65	
Ground protein isolate	58		1.58	27
Loblah bean	77	60		31
Lentils	45	30	0.93	51
Njugo bean	56			37
Peas	64	47	1.57	
Pigeon pea	57	52	1.54	37
Soybeans	73	61	2.32	47
Soy milk			2.10	55
Velvet bean	40	27		
Coconut	69		2.14	55
Cottonseed meal	67	53	2.25	47
Linseed	71	56	2.11	59
Pecan	60			61
Sesame	62	53	1.77	42
Sunflower seed	70	58	2.10	61
Alfalfa	57			50
Turnip green	52			33
Lupine	83			25
Beef	74	67	2.30	69
Chicken	74			64
Egg	94	94	3.92	100
Fish	76	80	3.55	71
Crustaceans	81			66
Molluscs	81			71
Fishmeal	81	65	3.42	60
Casein	80	72	2.86	58
Cow's milk	85	82	3.09	60
Brewer's yeast	67	56	2.24	57

[a]Values selected from an FAO report (1970) for which several different measures are available.

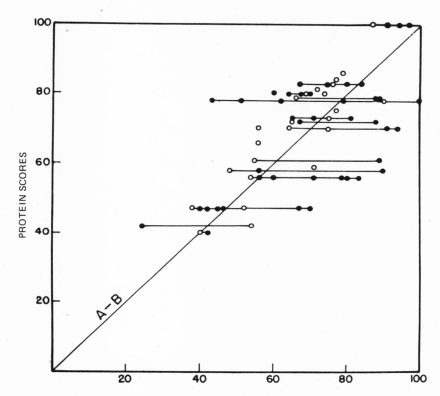

FIGURE 2 Taken from the 1957 FAO report on protein requirements. The committee assumed that an Amino Acid Score of zero should yield a BV of zero.

score should be capable of meeting protein needs if they were fed in
amounts 2 1/2 times greater than that required with egg protein (BV =
100).

Comparison of NPU and Amino Acid Score values taken from Table 1
shows essentially the same relationship (Figure 4), although with somewhat less deviation from expectation. According to this plot, a protein
of zero score yields an NPU of approximately 25 percent. If NPU is accepted as the true measure of protein quality, this implies that protein
needs can be met by feeding proteins of zero score at four times the
minimal requirement.

The hazards of collecting values from a widely scattered literature in
which the accuracy of neither the biological determination nor the
amino acid analysis is known, is, of course, well recognized. However,

this does not negate the clear fact that Amino Acid Score does not measure the same thing as NPU and BV.

It can be pointed out, of course, that when one is concerned with diets in which protein quality is reasonably high—NPU, BV, or Amino Acid Score above 60 or 70, for example—the error in the correction will be relatively small, regardless of which measure of protein quality is used. However, it is with diets of poor quality that correction is of immediate practical importance; for these the significance of the various measures of protein quality is in doubt.

The reasons for the discrepancy between the theoretical prediction and the experimental facts are now becoming clear. The basic assumption that has been made is that equivalent degrees of deficiency of any essential amino acid will produce the same metabolic effect. That is, a protein that supplies only 50 percent of the requirement of any essential amino acid should depress protein synthesis to a comparable degree and yield the same BV as a protein that supplies 50 percent of any other essential amino acid; and proteins that are completely lacking in any

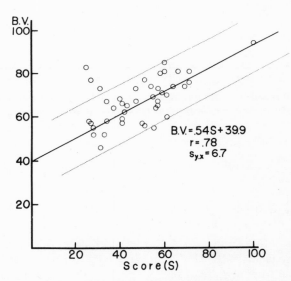

FIGURE 3 Data taken from Table 1. Amino Acid Score of zero apparently yields a BV of about 40 percent. The dotted lines define two standard errors on either side of the regression line. The three points in the upper-left-hand portion falling outside the dotted lines were omitted in calculating the regression line.

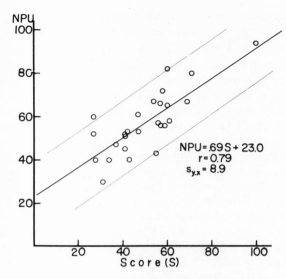

FIGURE 4 Relationship between NPU and Amino Acid Score from data in Table 1. Dotted lines define two standard errors on either side of the line.

essential amino acid should not allow any protein synthesis. They should be nutritionally equivalent to a diet containing no protein and should yield a BV of zero. It is now certain that these assumptions are not correct, although the details of the relationship remain to be thoroughly studied.

Bender (1961), for example, observed that a diet containing no lysine yielded an NPU of approximately 40 percent, rather than the expected value of zero. Diets lacking other essential amino acids yielded lower values, but rarely the expected zero. Similarly, Said and Hegsted (1970) found that adult rats fed a lysine-free diet lost weight slowly compared to rats fed diets lacking other essential amino acids. Only when the rats consumed diets lacking threonine, isoleucine, or sulfur-containing amino acids were the weight losses comparable to that obtained with a protein-free diet. Thus, with absolute deficiencies of most essential amino acids, the loss of tissue protein is less than would be predicted from Amino Acid Score.

Although the biochemical mechanisms are unknown, the logical explanation of these effects is that the body is able to adapt to some degree to amino acid deficiencies by decreasing the rates of catabolism of the appropriate amino acid and conserving it for reutilization. This adaptive response appears to be maximal for lysine, less so for most other amino acids, and practically nonexistent for threonine and a few others. Thus, we assume that, when lysine becomes limiting, its catabolism is greatly depressed, allowing the body to conserve it and, in turn, to conserve tissue protein. This results in a BV or NPU that is substantially above that predicted from the Amino Acid Score.

Hegsted and Chang (1965a,b), using young rats, proposed a slope–ratio assay for the estimation of protein quality. They showed, contrary to the assumptions of Miller and Payne (1961a,b), that over a rather wide range of protein intakes, up to those that produced near-maximum rates of gain, a linear relationship existed between body nitrogen and protein consumed. The slope of the line, relating protein consumed to body nitrogen when lactalbumin or egg protein was fed, was such that nearly 100 percent of the nitrogen consumed could be accounted for in tissue protein. They thus concluded that lactalbumin was an adequate standard and that the relative nutritive values of other proteins could be estimated by comparing the slope of its dose–response line for it to those obtained for other proteins under the same conditions.

It should be noted that the estimation of BV or NPU is, in fact, a slope–ratio assay based upon only two points—zero intake and some arbitrarily selected intake. The difference in response (nitrogen balance or retained nitrogen), divided by the difference in intake, defines the slope of a straight line connecting these two points. Thus, in the calculation of BV or NPU one must assume a linear response to increased dosage, although this cannot, in fact, be verified if only two points are available. Statistical evaluation of many assays by Hegsted and Worcester (1966) and Hegsted *et al.* (1968) demonstrated that, contrary to the assumptions made in the calculation of BV, NPU, or the slope–ratio assay, the dose–response lines often did not have a common intercept at zero dosage. Although it is sometimes difficult to demonstrate this with certainty when young rats are used in the assay, it is readily demonstrable in data obtained from studies with adult animals. Figure 5 shows the response of adult rats to varying intakes of lactalbumin and wheat gluten (Said and Hegsted, 1969). Whereas lactalbumin approximately fits the theoretical expectation that the response should be linearly related to protein eaten, wheat gluten obviously does not.

Inoue *et al.* (in press) have confirmed these kinds of findings. Figure 6 shows the nitrogen balance of rats fed varying intakes of lactalbumin and of wheat gluten. At low intakes, the authors could not distinguish between the BV of egg protein and of wheat gluten. In retrospect, it seems clear that, at intakes of wheat gluten approximating those required for maintenance of nitrogen balance or for weight maintenance, lysine is the limiting amino acid. However, at lower intakes, lysine is conserved and other factors, perhaps total nitrogen, become limiting.

Inoue and his associates have also demonstrated this phenomenon in human subjects. Figure 7 shows the nitrogen balance in young adult men fed varying levels of egg protein and wheat gluten. At intakes below 30 mg of N/kg, egg protein and wheat gluten were equally efficient and gave the same BV.

Examination of the dose–response lines (Figures 5, 6, and 7) clearly shows that the BV or NPU of egg protein or lactalbumin is approximately constant over a considerable range of intakes. On the other hand, the values for wheat gluten are entirely dependent upon the level fed. It must be concluded that neither of these methods appropriately

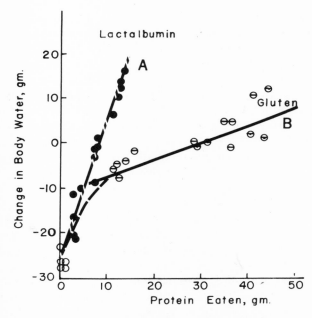

FIGURE 5 Dose-response lines in adult animals fed varying levels of lactalbumin and wheat gluten. Animals fed lactalbumin respond approximately according to theoretical expectation; the line is linear over the entire range tested. Since NPU is the slope of the line, it is approximately constant and characteristic of the protein. On the other hand, it is not linear at low intakes; and NPU will be variable and dependent upon the level fed. Data from Said and Hegsted (1969).

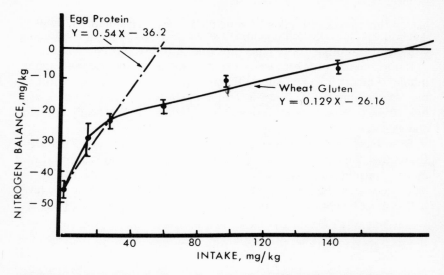

FIGURE 6 Nitrogen balance in adult rats fed varying levels of lactalbumin and wheat gluten. The response approximates that shown in Figures 5 and 6. Data from Inoue *et al*. (in press).

defines the nutritive value of wheat gluten or of most proteins. It is obviously incorrect to conclude that a protein that is free of lysine has a higher nutritive value than a protein that is free of any other essential amino acid.

The true significance of this adaptation to inadequate intakes of amino acids is not entirely clear. Although such adaptation should provide some protection against acute deficiencies, it may not be helpful in the long run. The fact that the response to gluten shown in Figures 5, 6, and 7 is linear over a substantial range on either side of nitrogen balance suggests that adaptation does not become operative until deficiency is rather severe. In any event, the linear portion of the dose–response line in these figures defines the efficiency of the protein to cover nitrogen needs in the range around nitrogen balance, the range of practical interest. Thus, in Figure 7, for example, the efficiency of wheat gluten is only 0.129, or 12.9 percent, in meeting the protein needs of man. Higher and variable values will be obtained if one calculates the slopes of lines between zero intake and any other selected intake, as in the calculation of BV or NPU.

Therefore, the most appropriate estimate of the nutritive value of a protein, whether estimated by nitrogen retention or by measures of growth in young animals, is probably the slope of the dose–response line—within ranges of intakes in which it is linear. This method, however, requires measurement at more than one level of protein intake and should not be calculated relative to zero intake.

OTHER METHODS OF ESTIMATING PROTEIN QUALITY

Protein Efficiency Ratio (PER)

Qualitative differences in protein quality can be demonstrated by many methods. The Protein Efficiency Ratio (PER), because of its relative simplicity, has been the method most widely used. Osborne *et al.* (1919) observed that young rats fed certain poor-quality proteins gained little weight and ate little protein, whereas those that were fed better-quality proteins gained more weight and consumed more protein. In an attempt to compensate for the difference in food intake, the investigators calculated the weight gain per gram of protein eaten, which has come to be called PER. The PER for any protein is dependent upon the amount of protein incorporated in the test diet. Therefore, a set of standardized conditions has been proposed (Derse, 1962). The use of 10 weanling

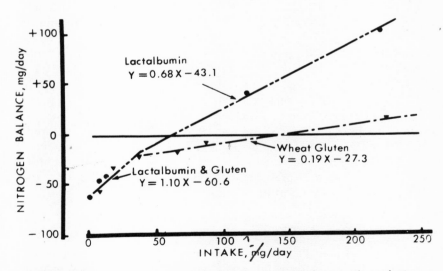

FIGURE 7 Nitrogen balance in human subjects fed varying levels of egg protein or wheat gluten. The response to egg protein is approximately linear but, as in Figure 5, the response to wheat gluten is not. Data of Inoue *et al.* (in press).

rats per test group, diets containing 9.09 percent protein (N × 6.25),
a test period of 4 weeks' duration, and the inclusion in each experiment
of a group receiving statdardized casein. PER is calculated as the average
total weight gain divided by the average amount (grams) of protein consumed. Since PER, as determined in various laboratories, was not constant for the same protein, it was recommended that a corrected value
be calculated on the basis of an assumed PER of the standardized casein
of 2.50 (corrected PER = 2.50 × PER/PER of reference casein).

In spite of its simplicity, the PER has been severely criticized as a
measure of protein quality (Mitchell, 1944; Hegsted and Worcester,
1947; Hegsted and Chang, 1965a). The most common criticisms are
that dietary protein required for the maintenance of the animal is not
credited in the measurement of PER, and that variation may occur in
body composition so that PER is not afford an adequate measure of
nitrogen retention. From a theoretical point of view, the major criticism of PER is that it is not a direct function of the nutritive value of
the protein; it is influenced by the weight gain, the amount of food
consumed, the amount of protein in the diet, and the nutritive quality
of the protein in the diet. The relationship between these factors is
complex and undefined. PER also has the disadvantage that even under
standardized conditions it is not reproducible in different laboratories
(Derse, 1962). It is of interest that in a collaborative study, corrected
PER values showed larger differences among laboratories than did un-
corrected values, indicating that correction was not appropriate and of
no advantage.

Clearly, PER is not proportional to the nutritive quality of proteins
tested. A protein that has a PER of 1.5, for example, cannot necessarily
be assumed to have 50 percent of the value of a protein showing a PER
of 3. Thus, a statement that "the total protein (must have) . . . a Biological Value (as defined by PER) not less than 70 percent of casein,"
such as has been proposed (Edwards, 1970) as a standard for textured
protein products, is not meaningful. A judgment often can be made,
based on PER, that one protein is better or worse than another; but it
is not appropriate to express these differences as percentages, since the
differences are not proportional to nutritional quality.

Net Protein Ratio (NPR)

A major criticism of PER has been that it does not take into account
the protein required for maintenance, since only gain in weight is used
in the calculation. Bender and Doell (1957) suggested that this difficulty could be avoided by including in each test a group of animals fed

a protein-free diet. Net Protein Ratio (NPR) was then calculated as the overall difference in gain (gain in weight of the test group plus loss in weight of the protein-free group) divided by the amount of protein eaten. It is apparent that, if body composition is constant, this procedure is identical to NPU except that it is expressed in arbitrary units that are less useful than the percentage of protein utilized.

Relative Nutritive Value (RNV)

Hegsted and associates (Hegsted and Chang, 1965a,b; Hegsted and Worcester, 1966; Hegsted et al., 1968; Hegsted and Neff, 1970) proposed a slope–ratio assay in which the slope of the regression line relating body protein (or body water) of rats fed a standard protein (egg protein or lactalbumin)—assumed to have maximal nutritive value—was compared to that of rats fed the test protein. In developing this method, the assumption was made, as in the estimation of BV and NPU, that the dose–response lines should have a common intercept at zero dose. However, by using several levels of intake, this assumption can be statistically tested; and it can be shown that the regression lines for many, perhaps most, proteins do not have a common intercept. Thus, the general statistical procedures used for the evaluation of slope–ratio assays of this kind are not entirely valid. As has been indicated in Figures 5–7 and previously discussed, the deviation from theoretical expectation will depend upon the specific protein under test. When young rats are fed diets of varying protein content, up to levels that allow substantial increases in body weight, the slope of the regression line may not be greatly influenced, whether or not one forces the regression through a common zero point. However, since it is certain that the regression lines do not meet at a common point, a slope–ratio assay is not entirely appropriate. Rather, the slope of the regression line relating dose to response should be calculated for each protein individually and then compared to that of the standard protein.

Nitrogen Balance Index

Allison and Anderson (1945) showed, as noted above, that Biological Value, in an ideal situation at least, is the slope of the regression line relating nitrogen balance to nitrogen intake. They also suggested that the determination of the slope of such a line, as defined by several points, might be preferable to the ordinary determination of BV. This concept is valid, provided that the nitrogen balance is determined in the region of intake in which the balance is linearly related to intake. The slopes of the linear regression lines shown in Figures 6 and 7 would be measures of Nitrogen Balance Index.

COMMENT ON ESTIMATING PROTEIN QUALITY

It should be clear that, in an ideal situation in which the change in retained nitrogen is linearly related to intake, as is approximated by the regression lines shown for lactalbumin or egg protein in Figures 6 and 7, Biological Value, Net Protein Utilization, Net Protein Ratio, Relative Nutritive Value, and Nitrogen Balance Index measure the same thing. Biological Value and Nitrogen Balance Index would be slightly larger than the others, since these are calculated on absorbed nitrogen rather than on protein consumed. However, for most proteins the lines are not linear. Thus, measurements of BV, NPU, and NPR will lead to overestimates of the nutritive value of such proteins. Relative Nutritive Value and Nitrogen Balance Index are assumed to give correct values if the proteins are tested at appropriate intakes, i.e., those in the linear part of the dose–response curve. Only if several levels of protein are tested can one determine whether or not the dose–response curve is linear. Thus an adequate assay requires the use of multiple doses. The result obtained at zero intake of protein should not be included in the calculation of the slope unless the entire dose–response curve is linear.

Adult versus Young Animals

Further uncertainties concerning the meaning of measurements of nutritive value of proteins have recently surfaced. As noted earlier, evidence suggests that the essential amino acid requirements of adult man are considerably lower per unit of protein required than are those of infants and young children. This observation indicates that many proteins should be used more efficiently in meeting the requirements of adults than those of children. Similar observations repeatedly appear in the literature in studies on adult and young rats (Osborne and Mendel, 1916, 1919; Burroughs *et al.*, 1940), indicating that various proteins are more adequate in meeting needs for maintenance than for growth. Comparisons of the estimates of the needs for essential amino acids of growing rats (Rama Rao *et al.*, 1959) and adult rats (Said and Hegsted, 1970) also support this conclusion.

However, observations such as those presented by Clark *et al.* (1967) and Romo and Linkswiler (1969) are not entirely consistent with these conclusions. These authors observed improvements in nutritive value of diets that contained essential amino acids in excess of the apparent requirements. Considerably more information about the nutritive value of various proteins in adult man and adult animals, obtained by more acceptable methods, are needed.

A second uncertainty stems from the observation of Calloway and Margen (1971) that egg protein is relatively poorly utilized in adult man as compared to the young rat. They concluded that it was only about 65 percent efficient, rather than 90–95 percent, as has been repeatedly demonstrated in young rats. Confirmatory evidence is indicated by the the data of Inoue *et al.* (in press), as shown in Figures 6 and 7; the efficiency of utilization by adult rats and man of both lactalbumin and egg protein is substantially less than the values reported for young rats. It is of some interest that the estimated efficiency of utilization of lactalbumin in growing cebus monkeys (Samonds and Hegsted, 1973) is also between 60–70 percent. This observation suggests that there may be differences among species as well as differences that depend upon age in protein utilization.

It is possible that adult animals that have been rather seriously depleted of protein utilize it much more efficiently than do those that are only moderately depleted and, under these conditions, may approximate the utilization observed in young rats. This aspect of the problem deserves more adequate study.

Tissue Regeneration

A variety of techniques based on recovery of weight or of specific tissues after protein depletion have been proposed (Whipple and Robscheit-Robbins, 1925; Pearson *et al.*, 1937; Campbell and Kosterlitz, 1948; Goyco, 1956). The specific merits of such assays, as opposed to weight gain of young rats, for example, remain to be demonstrated.

Microbiological Assays

Many microorganisms require the amino acids that are essential for monogastric animals. If it were possible to find microorganisms that required amino acids not only in the same pattern, but also in the same relative amounts, their growth response when supplied with limited amounts of various proteins or protein hydrolysates would provide a simple and efficient assay of nutritive value. Considerable effort has been directed toward this end (Horn and Warren, 1961, 1964; Ford, 1964; Baum and Haenel, 1965), and the results obtained with some organisms resemble those observed in some of the rat assays described. The limitations of animal assays, however, mean that they may provide an inadequate base for comparison.

Plasma Amino Acids

As already noted, changes in plasma amino acid concentrations after the feeding of various proteins can, under certain conditions, yield estimates of the nutritional quality of the proteins fed. However, the range in concentration of each of the amino acids in plasma of normal animals is relatively large. This variability imposes serious limitations upon the quantitative interpretation of any changes in the levels observed. Thus, while it may be possible to identify the limiting amino acid in certain proteins by this technique, the likelihood that good quantitative assays for nutritional quality can be developed, using plasma amino acid concentrations, is not promising.

EFFECTS OF DIGESTION AND AVAILABILITY OF AMINO ACIDS

If significant quantities of nitrogen or essential amino acids are not absorbed from the gut when a protein is fed, the nutritional quality of that protein will be impaired. Much evidence suggests that, in the case of many proteins, digestibility is not a primary factor in determining nutritive value; but detailed information on the subject is insufficient. For example, if a protein is 95 percent digested, the 5 percent not absorbed would be of little significance if the nonabsorbed portion were of the same composition as the total protein. If, however, this relatively small fraction contained primarily one or two amino acids, the effect might be substantial.

Such information as is available upon the amino acid content of feces from animals fed various proteins does not indicate that the digestive or absorptive processes are very selective. However, it has been shown that large amounts of amino acids are secreted into the gut (Nasset, 1964, 1968) and that the actual source of the amino acids that eventually appear in the feces is unknown. Neither the destructive nor synthetic capabilities of the gastrointestinal flora appear to have been investigated. Thus, while data on amounts of amino acids in the feces may be useful in comparative investigations in which different proteins are studied, these values cannot be considered an absolute measure of lack of digestion.

The formula shown for the determination of "true digestibility" is based upon the assumption that F_o, the nitrogen in the feces when a protein-free diet is fed, is constant and independent of the diet. There is considerable evidence to show that this is not strictly true and that F_o varies with the kind and amount of food consumed (Mitchell, 1924a, 1942; Weinstein *et al.*, 1961). Although the "bulk" and "roughage" in a diet have often been assumed to be major factors affecting fecal nitrogen excretion, direct studies (McCance and Widdowson, 1947; McCance and Walsham, 1948; Dreyer, 1969) do not confirm these assumptions. Because a large proportion of the feces is composed of bacteria and because relatively little is known of the factors that control the intestinal flora (Weinstein *et al.*, 1961), the basic concepts underlying the assumptions made in the determination of digestibility are subject to question. One might suppose that, when relatively pure proteins are compared, the remainder of the diet being held essentially constant, valid assessments could be made. However, when crude food materials are tested, a multitude of factors may influence the amount and kind of fecal flora, the rate of transit through the gut, the excretions into the gut, etc.; and digestibility of proteins should probably not be considered a function of the protein fed. Presumably, true values for digestion of proteins might be obtained by the use of isotopically labeled food proteins. However, since the "amino acid pool" is rapidly labeled with dietary proteins, which then appear in the gastrointestinal secretion, even this approach has serious limitations.

In processed foods, particularly those subjected to heat treatment, reactions may occur that result in the destruction of amino acids or the formation of compounds that are not easily digestible or utilized by the mammalian organism (FNB, 1950; Liener, 1958; Bjarnason and Carpenter, 1970). These kinds of reactions apparently involve primarily lysine, methionine, arginine, histidine, and tryptophan. "Unavailable lysine" has been particularly studied, especially by Carpenter and co-workers (Carpenter, 1960). They have used a method that involves the reaction of the terminal amino group in lysine with 1-fluoro-2,4-dinitrobenzene. During heat treatment this amino group may react with carbohydrates, or possibly other materials, and is then unavailable to react with fluorodinitrobenzene. In some tests (Carpenter *et al.*, 1963), the available lysine determined by the chemical method correlated reasonably well with available lysine determined by animal assay, but in other studies (Boctor and Harper, 1968) results were less encouraging. Comparable chemical methods for other amino acids are not available. Microbiologic assays (Ford, 1964; Horn and Warren, 1964) have also been used to assess the "availability" of various amino acids. The significance of all such methods clearly depends upon evidence that they do indeed reflect availability of the amino acids to animals. Essentially no evidence available relates the results of applying these kinds of methods on "available amino acids" to human nutrition.

Various *in vitro* methods (Mauron *et al.*, 1955; Sheffner *et al.*, 1956) have been proposed to evaluate digestibility. While large differences in the rate and extent of release of amino acids from different proteins can be shown, the relationship between such findings and "true digestibility" in the gastrointestinal tract is relatively unexplored.

There is an abundant literature (Liener, 1958, 1966; Birk, 1968) demonstrating poor digestibility and other untoward effects when raw legumes of various kinds are fed to animals. These products contain heat labile "trypsin inhibitors" and hemagglutinins. However, it is not certain that these factors explain all of the diverse effects observed, e.g., hypertrophy of the pancreas and increased need for methionine. The significance of these effects in human nutrition is relatively unknown; because legumes are usually cooked for man's consumption, they may be of minimal importance.

CONCLUSION

A critical examination of the various methods in use for the evaluation of protein quality shows that the situation is less satisfactory than has been widely assumed. Although BV and NPU have been generally thought to be the methods of choice, evidence has shown that these methods overestimate the nutritional quality of many proteins. This outcome results from the ability of the organism to adapt, in varying degree, to different amino acid deficiencies. It seems particularly clear at this time that proteins limiting in lysine, which include most, if not all, of the cereal proteins, yield BV's and NPU's that are in excess of prediction based on Amino Acid Score.

The best estimate of protein quality, whether obtained by nitrogen balance or by growth responses, will be obtained from the slope of the dose–response line in the range of intakes where the line is linear. This requires that tests be run at several protein intakes, and the response should not be calculated from zero intake.

Considerable data are now available showing that measurements of the quality of many proteins will be higher in adult animals than in growing animals and that the efficiency of protein utilization may be considerably lower in man than in the growing rat. Species differences, as well as the effects of age or growth rate, may be involved.

Amino Acid Score may be the most logical method of predicting protein quality, provided amino acid availability or indigestibility is not a determining factor. However, confirmation of this supposition will require the collection of appropriate information by more precise biological methods.

REFERENCES

Allison, J. B., and J. A. Anderson. 1945. The relationship between absorbed nitrogen, nitrogen balance and biological value of proteins in adult dogs. J. Nutr. 29: 413–420.

Baum, F., and H. Haenel. 1965. Estimation of biological value of protein with *Tetrahymina pyriformes* W. I. Method and estimation of the value of some foods. Nahrung 9:517–525.

Bender, A. E. 1961. Determination of the nutritive value of proteins by chemical analysis. Pages 407–424 *in* Progress in meeting protein needs of infants and preschool children. Publ. 843. National Academy of Sciences, Washington, D.C.

Bender, A. E., and B. H. Doell. 1957. Biological evaluation of proteins: A new aspect. Br. J. Nutr. 11:140–148.

Birk, Y. 1968. Chemistry and nutritional significance of proteinase inhibitors from plant sources. Annu. N.Y. Acad. Sci. 146:388–399.

Bjarnason, J., and K. J. Carpenter. 1970. Mechanisms of heat damage in proteins. II. Chemical changes in pure proteins. Br. J. Nutr. 24:313–329.

Block, R. J., and H. H. Mitchell. 1946–47. The correlation of the amino acid composition of proteins with their nutritive value. Nutr. Abstr. Rev. 16:249–278.

Boctor, A. M., and A. E. Harper. 1968. Measurement of available lysine in heated and unheated foodstuffs by chemical and biological methods. J. Nutr. 94:289–296.

Brush, M., W. Willman, and P. P. Swanson. 1947. Amino acids in nitrogen metabolism with particular reference to the role of methionine. J. Nutr. 33:389–410.

Burroughs, E. W., H. S. Burroughs, and H. H. Mitchell. 1940. The amino acids required for the complete replacement of endogenous losses in the adult rat. J. Nutr. 19:363–384.

Calloway, D. H., and S. Margen. 1971. Variation on endogenous nitrogen excretion and dietary nitrogen utilization as determinants of human protein requirement. J. Nutr. 101:205–216.

Campbell, R. M., and H. W. Kosterlitz. 1948. The assay of the nutritive value of a protein by its effect on liver cytoplasm. J. Physiol. 107:383–398.

Carpenter, K. J. 1960. The estimation of the available lysine in animal-protein foods. Biochem. J. 77:604–610.

Carpenter, K. J., B. E. March, C. K. Milner, and R. C. Campbell. 1963. A growth assay with chicks for the lysine content of protein concentrates. Br. J. Nutr. 17:309–323.

Clark, H. E., K. Fugate, and P. E. Allen. 1967. Effect of four multiples of a basic mixture of essential amino acids on nitrogen retention of adult human subjects. Am. J. Clin. Nutr. 20:233–242.

Cox, W. M., A. J. Mueller, R. Elman, A. A. Albanese, K. S. Kemmerer, R. W. Barton, and L. E. Holt, Jr. 1947. Nitrogen retention studies on rats, dogs and man: The effect of adding methionine to an enzymic casein hydrolysate. J. Nutr. 33:437–457.

Derse, P. H. 1962. Evaluation of protein quality (biological method). J. Assoc. Off. Anal. Chem. 45:418–422.

Dreyer, J. J. 1957. The nitrogen:water ratios of albino rats and their use in protein-evaluation tests. Br. J. Nutr. 11:22–27.

Dreyer, J. J. 1962. The estimation of (1) lean body weight from carcass water content and (2) mean lean body weight from nitrogen retention in rats used in nitrogen balance experiments. Proc. Nutr. Soc. S. Afr. 3:163–168.

Dreyer, J. J. 1969. The biological assessment of protein quality: Effects of consumption of "crude fiber," NaCl, and body hair on faecal nitrogen excretion in the cat. S. Afr. Med. J. 43:776–786.

Edwards, C. C. 1970. Textured protein products. Proposed standard of identity. Fed. Regist. 35, No. 236. December 5, 1970.

FAO (Food and Agriculture Organization). 1957. Protein requirements. FAO Nutr. Stud. Ser. No. 16. FAO, Rome. 52 pp.

FAO (Food and Agriculture Organization). 1970. Amino-acid content of foods and biological data on proteins. FAO Nutr. Stud. No. 24. FAO, Rome. 285 pp.

FAO/WHO (Food and Agriculture Organization/World Health Organization). 1965. Protein requirements. WHO Tech. Rep. Ser. No. 301; FAO Nutr. Meet. Rep. Ser. No. 37. WHO, Geneva. 71 pp.

FNB (Food and Nutrition Board, National Research Council). 1950. The problem of heat injury to dietary protein. Repr. Circ. Ser. No. 131. National Academy of Sciences, Washington, D.C. 19 pp.

FNB (Food and Nutrition Board, National Research Council). 1959. Evaluation of protein nutrition. Publ. 711. National Academy of Sciences, Washington, D.C. 61 pp.

Ford, J. E. 1964. A microbiological method for assessing the nutritional value of proteins. III. Further studies on the measurement of available amino acids. Br. J. Nutr. 18:449–460.

Goyco, J. A. 1956. A study of the relation between the liver protein regeneration capacity and the hepatic necrogenic activity of yeast proteins. J. Nutr. 58:299–308.

Hegsted, D. M., and Y. Chang. 1965a. Protein utilization in growing rats. I. Relative growth index as a bioassy procedure. J. Nutr. 85:159–168.

Hegsted, D. M., and Y. Chang. 1965b. Protein utilization in growing rats at different levels of intake. J. Nutr. 87:19–25.

Hegsted, D. M., and R. Neff. 1970. Efficiency of protein utilization in young rats at various levels of intake. J. Nutr. 100:1173–1179.

Hegsted, D. M., and J. Worcester. 1947. A study of the relation between protein efficiency and gain in weight on diets of constant protein content. J. Nutr. 33:685–702.

Hegsted, D. M., and J. Worcester. 1966. Assessment of protein quality with young rats. In Proc. 7th Int. Congr. Nutr. Vol. 4. Vieweg and Sohn, Braunschweig, West Germany.

Hegsted, D. M., R. Neff, and J. Worcester. 1968. Determination of the relative nutritive value of proteins. Factors affecting precision and validity. J. Agric. Food Chem. 16:190–195.

Henry, K. M., and J. Toothill. 1962. A comparison of the body-water and nitrogen balance-sheet methods for determining the nutritive value of proteins. Br. J. Nutr. 16:125–133.

Horn, M. J., and H. W. Warren. 1961. Availability of amino acids to microorganisms. III. Development of a method for comparison of hydrolysates of foods with synthetic simulated mixtures. J. Nutr. 74:226–232.

Horn, M. J., and H. W. Warren. 1964. Availability of amino acids to microorganisms. IV. Comparison of hydrolysates of lactalbumin, oatmeal and peanut butter with simulated amino acid mixtures by growth response of microorganisms. J. Nutr. 83:267–272.

Inoue, G., Y. Fujita, K. Kishi, and Y. Niiyama. (In press.) Nutritive values of egg protein and wheat gluten in young men. Proc. 9th Int. Congr. Nutr.

Johnson, R. M., J. J. Deuel, M. G. Morehouse, and J. W. Mehl. 1947. The effect of methionine upon the urinary nitrogen in men at normal and low levels of protein intake. J. Nutr. 33:371–387.

Liener, I. E. 1958. Effect of heat on plant proteins. Pages 79–129 in A. M. Altschul, ed. Processed plant protein foodstuffs. Academic Press, New York.

Liener, I. E. 1966. Toxic substances associated with seed proteins. Pages 178–194 in World protein resources. Adv. Chem. Ser. 57. American Chemical Society, Washington, D.C.

McCance, R. A., and C. M. Walsham. 1948. Digestibility and absorption of calories, proteins, purines, fat and calcium in wholemeal wheaten bread. Br. J. Nutr. 2: 26–41.

McCance, R. A., and E. M. Widdowson. 1947. Digestibility of English and Canadian wheats with special reference to the digestibility of wheat protein in man. J. Hyg. 45:59–64.

Mauron, J., F. Mottu, E. Bujard, and R. H. Egli. 1955. The availability of lysine, methionine, and tryptophan in condensed milk and milk powder. In vitro digestion studies. Arch. Biochem. Biophys. 59:433–451.

Miller, D. S. 1963. A procedure for determination of NPU using rats body N technique. Pages 34–37 in Evaluation of protein quality. Publ. 1100. National Academy of Sciences, Washington, D.C.

Miller, D. S., and A. E. Bender. 1955. The determination of the net utilization of proteins by a shortened method. Br. J. Nutr. 9:382–388.

Miller, D. S., and P. R. Payne. 1961a. Problems in the prediction of protein values of diets. Br. J. Nutr. 15:11–19.

Miller, D. S., and P. R. Payne. 1961b. Problems in the prediction of protein values of diets: The use of food composition tables. J. Nutr. 74:413–419.

Mitchell, H. H. 1924a. The biological value of proteins at different levels of intake. J. Biol. Chem. 58:905–922.

Mitchell, H. H. 1924b. A method of determining the biological value of protein. J. Biol. Chem. 58:873–903.

Mitchell, H. H. 1942. An evaluation of feeds on the basis of digestible and metabolizable nutrients. J. Anim. Sci. 1:159–173.

Mitchell, H. H. 1944. Determination of the nutritive value of the proteins of food products. Ind. Eng. Chem. (analytical edition) 16:696–700.

Mitchell, H. H., and G. G. Carman. 1926. The biological value of the nitrogen mixtures of patent white flour and animal foods. J. Biol. Chem. 68:183–215.

Mitchell, H. H., J. R. Beadles, and J. H. Kruger. 1927. The relation of the connective tissue content of meat to its protein value in nutrition. J. Biol. Chem. 73: 767–774.

Nasset, E. S. 1964. Role of the digestive tract in protein metabolism. Am. J. Dig. Dis. 9:175–190.

Nasset, E. S. 1968. Contribution of the digestive system to the amino acid pool. Pages 80–87 *in* J. H. Leathem, ed. Protein nutrition and free amino acid patterns. Rutgers University Press, New Brunswick, N.J.

Osborne, T. B., and L. B. Mendel. 1916. The amino acid minimum for maintenance and growth, as exemplified by further experiments with lysine and tryptophane. J. Biol. Chem. 25:1–12.

Osborne, T. B., and L. B. Mendel. 1919. The nutritive value of the wheat kernel and its milling products. J. Biol. Chem. 37:557–601.

Osborne, T. B., L. B. Mendel, and E. L. Ferry. 1919. A method of expressing numerically the growth-promoting value of proteins. J. Biol. Chem. 37:223–229.

Pearson, P. B., C. A. Elvehjem, and E. B. Hart. 1937. The relation of protein to hemoglobin building. J. Biol. Chem. 119:749–763.

Rama Rao, P. B., V. C. Metta, and B. C. Johnson. 1959. The amino acid composition and the nutritive value of proteins. I. Essential amino acid requirements of the growing rat. J. Nutr. 69:387–391.

Romo, G. S., and H. Linkswiler. 1969. Effect of level and pattern of essential amino acids on nitrogen retention of adult man. J. Nutr. 97:147–153.

Said, A. K., and D. M. Hegsted. 1969. Evaluation of dietary protein quality in adult rats. J. Nutr. 99:474–480.

Said, A. K., and D. M. Hegsted. 1970. Response of adult rats to low dietary levels of essential amino acids. J. Nutr. 100:1363–1375.

Samonds, K. W., and D. M. Hegsted. 1973. Protein requirements of young cebus monkeys (*Cebus albifrons* and *apella*). Am. J. Clin. Nutr. 26:30–40.

Sheffner, A. L., G. A. Eckfeldt, and H. Spector. 1956. The pepsin–digest–residue (PDR) amino acid index of net protein utilization. J. Nutr. 60:105–120.

Thomas, K. 1909. Ueber die biologische Wertigkeit der Stickstoffsubstanzen in verschiedenen Nahrungsmitteln; Beiträge zur Frage nach dem physiologischen Stickstoffminimum. Arch. Physiol. 219–302.

Weinstein, L., R. E. Olson, T. B. Van Itallie, E. Cass, D. Johnson, and F. J. Ingelfinger. 1961. Diet as related to gastrointestinal function. Report of a Committee of the Council on Foods and Nutrition. J. Am. Med. Assoc. 176:935–942.

Whipple, G. H., and F. S. Robscheit-Robbins. 1925. Blood regeneration in severe anemia. I. Standard basal ration bread and experimental methods. Am. J. Physiol. 72:395–408.

J. M. McLAUGHLAN

Nutritional Significance of Alterations in Plasma Amino Acids and Serum Proteins

During digestion, proteins are hydrolyzed to free amino acids that are absorbed rapidly from the small intestine and pass, via the mesenteric and portal veins, to the liver before entering the general circulation. After a high-protein meal, free plasma amino acid (PAA) concentrations generally increase, the changes being greater in portal than in systemic blood (Denton and Elvehjem, 1954). After a meal containing a low or moderate amount of protein, systemic PAA changes may be insignificant (Wynne and Cott, 1956; Nasset, 1968).

Christensen (1964) has stated that "the plasma level of each of the various amino acids is unquestionably controlled by balances between the entry and exit from the plasma, as in the case for the blood sugar level. Under basal conditions we can expect each level to tend to reach a characteristic value, as does the blood sugar." He neither elucidated the homeostatic mechanisms involved nor gave evidence for their existence. The following observations on PAA concentrations, however, support his theory: (1) Animals in the fasting state tend to have a characteristic PAA pattern (Henderson et al., 1949); (2) certain hormones have antagonistic effects on those concentrations (Munro, 1964); (3) the protein of a single meal is diluted in the gut with a relatively large amount of endogenous protein, so that relatively minor alterations in PAA patterns may occur after ingestion of moderate amounts of proteins (Nasset, 1968); and (4) ingestion of a diet that has an amino acid imbalance brings about a marked fall in the plasma concentration of the limiting amino acid—the low PAA concentration triggers some mechanism that inhibits ingestion of food, which in turn tends to keep the pattern from becoming more abnormal (Harper and Rogers, 1965; Leung and Rogers, 1969).

FASTING PAA CONCENTRATIONS

Fasting concentrations for seven of the essential amino acids in four species of animals are shown in Figure 1. Except for the high lysine value in the rat plasma, patterns are remarkably similar for these widely different species of mammals (Potter *et al.*, 1968; Anderson and Linkswiler, 1969; Boomgaardt and McDonald, 1969). Patterns for the dog (Longenecker and Hause, 1959) and the guinea pig (McLaughlan, unpublished) are also similar to those in Figure 1. Fasting threonine values

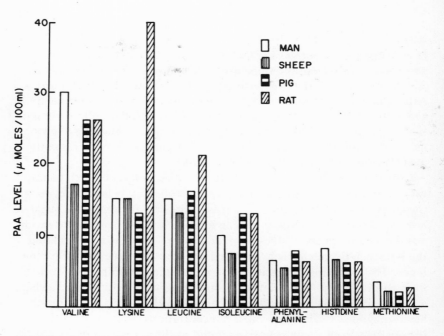

FIGURE 1   Concentrations of essential amino acids in plasma of four species of animals fasted for approximately 24 h. Data from Potter *et al.*, 1968; Anderson and Linkswiler, 1969; Boomgaardt and McDonald, 1969.

are not given, but these are influenced by previous diet in the rat and tend to be variable (McLaughlan, unpublished). Young *et al.* (1971) reported that plasma tryptophan concentrations in fasted adult humans reflected the tryptophan content of the diet.

Swendseid *et al.* (1969) observed PAA-concentration changes in obese humans starved for periods up to 117 days. Initially, these subjects tended to have higher-than-normal concentrations of branched-chain amino acids, which usually fell during starvation. However, despite the continuous loss of body nitrogen, PAA concentrations (excepting branched-chain) remained near the prefast values.

Clinical findings on kwashiorkor (Holt *et al.*, 1963) clearly indicated that fasting PAA concentrations may be drastically altered by prolonged ingestion of a protein-deficient diet. It has been shown, in the laboratory, that rats fed graded amounts of a balanced amino acid mixture evidenced a direct relationship between dietary level and fasting plasma amino acid concentration (Young and Zamora, 1968). Similarly, rats fed restricted amounts of a 10-percent protein diet (egg white) showed markedly lower plasma concentration of most of the essential amino acids other than histidine (McLaughlan, unpublished).

Unlike mammals, chicks do not have a characteristic fasting PAA pattern. The concentrations of most amino acids, particularly lysine, increase steadily in chicks fasted for periods up to 36 h (Hill and Olsen, 1963). The relatively stable PAA pattern in mammals fasted for 12–24 h makes it useful as a reference or baseline (zero time) in PAA studies. However, it has recently become clear that the concentrations of free amino acids in plasma fluctuate regularly on a biological time cycle (Feigin *et al.*, 1968; Wurtman, 1970). Consequently, amino acid rhythms should be taken into account in all studies involving PAA concentrations.

HORMONES

Protein metabolism is controlled by hormones. Insulin, growth hormone, and testosterone have anabolic effects leading to nitrogen retention, whereas thyroxine, ACTH, and corticosteroid hormones have the opposite effect (Munro, 1964). The injection of insulin into a fasting animal causes a rapid uptake of amino acids by muscle; the pattern of amino acids removed from the plasma apparently resembles the amino acid composition of muscle (Munro, 1964). Similarly, the pattern of amino acids apparently removed from the plasma of adult human subjects after oral administration of glucose resembled the amino acid composition of high-quality proteins (Swendseid *et al.*, 1967).

However, Knipfel *et al.* (1969) did not confirm this finding in studies with growing rats; after oral administration of glucose, lysine concentration fell very markedly, but the concentration of plasma tryptophan was unaltered. Manchester (1970) concluded that the uptake of threonine was not stimulated by insulin, although enhanced incorporation into protein occurred. The administration of leucine (orally), or intravenous infusion of any one of several amino acids, may produce hypoglycemia due to release of insulin. Glucagon causes a sustained decrease in PAA concentrations; and Bromer and Chance (1969) suggested that, while the main effect of glucagon may be on gluconeogenic enzymes, its early action may be on uptake by the tissues. The topic of hormones and free amino acids has been reviewed extensively by Munro (1970), who pointed out that, in addition to hormones controlling PAA concentrations, the relative amounts of certain amino acids in plasma may act as regulators for the secretion of insulin, glucagon, growth hormone, and corticosteroids.

DIETARY FACTORS

Earlier workers have reported that the increase of concentration of various amino acids in plasma after ingestion of differing high-protein meals was proportional to the amount of amino acids in the meal. Investigators were hopeful that a method based on PAA concentrations could provide a kind of *in vivo* chemical score that rated protein quality, pointed out which amino acid was limiting, and gave the relative order of abundance and availability of amino acids. The important study of Longenecker and Hause (1959), in particular, aroused interest, because it raised the possibility of being able to rate protein quality through an *in vivo* method that might be readily applicable to humans. It was soon discovered, however, that dietary factors other than protein altered PAA concentrations and that peak PAA concentrations could occur in portal blood within 30 min after ingestion of food (Guggenheim *et al.*, 1960). Stomach-emptying time, which varies for different proteins (Rogers *et al.*, 1960), is therefore important. Other workers demonstrated markedly different rates of absorption of different amino acids (Gitler, 1964).

As noted above, Nasset (1968) reported the dilution of dietary proteins of a single test meal by relatively large amounts of endogenous protein. As a result of these findings, some workers have questioned the value of PAA concentrations in nutritional studies. One finding that stands out clearly from the many PAA studies is that the concentration of the limiting amino acid tended to fall, often markedly, in plasma of animals that were fed deficient diets for 1 day or more. This fall in PAA concentration may be transient, however, and the concentration may rise soon after an animal begins to fast (McLaughlan *et al.*, 1966). A few workers (Wynne and Cott, 1956; Nasset, 1968) reported only minor changes in PAA concentrations after single test meals. It is clear from the studies of Longenecker and Hause (1959) that, in order to overcome homeostatic mechanisms and bring about meaningful PAA changes, animals must consume some minimum amount of protein and that blood must be sampled at the appropriate time if PAA changes are to be detected.

The importance of some of these factors is illustrated in Figure 2. In this instance, weanling rats that had been fasted for approximately 12 h were given wheat gluten diets; PAA concentrations were measured at regular intervals. The data are expressed in Figure 2 as PAA scores so that response curves for all amino acids could be plotted on the same scale. PAA changes were erratic when the dietary protein level was 5 percent or 10 percent and 1 g of diet was fed. Under these conditions, lysine, although known to be limiting, could not be so demonstrated. When the dietary protein level was raised to 20 percent, however, plasma lysine concentration decreased to a minimum value 3 h post-prandial and then rose toward the fasting value (PAA score of 100). A similar, though more severe, depression in plasma lysine concentration occurred when 1 g of diet containing 40 percent protein was fed. Other amino acids showed fluctuations in plasma concentrations but tended to rise as the protein level of the diet increased.

When 4 g of diet containing 5 percent protein was fed, plasma concentrations of both lysine and threonine decreased continuously for 6 h after the meal. When the protein content of the meals was increased, however, the fact that lysine was the limiting amino acid became more and more obvious as the difference in its behavior, compared to that of other amino acids, became progressively greater. Although the effect of level of protein was very marked, it was apparent also that the quantity of diet ingested was important in making the limiting amino acid stand out clearly in the PAA pattern.

In similar studies, Anderson *et al.* (1968) observed PAA concentrations, serine–threonine dehydratase activity, and food consumption in rats undergoing adaptation to high-protein diets. After one day, PAA concentrations were high and food consumption was low. During the next few days, PAA concentrations dropped, while liver serine–threonine dehydratase and food consumption increased.

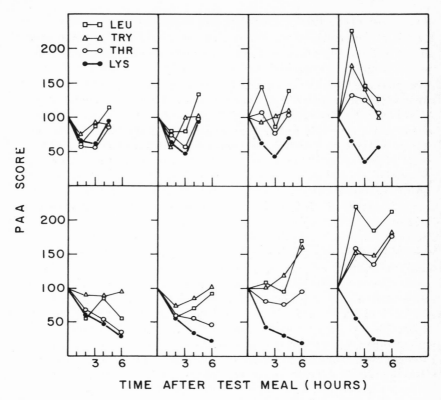

FIGURE 2 Interrelationships among quantity of protein ingested, level of dietary protein, time after meal, and PAA scores (PAA score = mean PAA concentration of test groups over that of fasted group X 100). Data given in the upper frames were derived from rats given 1 g of wheat gluten diet; data in the lower frames are from rats given 4 g of wheat gluten diet. Wheat gluten diets contained 5, 10, 20, and 40 percent protein (from left to right). Source: J. M. McLaughlan and F. J. Noel, unpublished data.

Snyderman *et al.* (1970) reported that methionine and tyrosine accumulated markedly in plasma of premature infants who were fed a high level of cow's milk protein (9 g/kg). They felt that the elevation of tyrosine was a consequence of a maturational defect in the metabolism of tyrosine at the level of parahydroxyphenyl pyruvic acid. Similarly, the high methioine concentration may result from a deficiency of liver cystathionase in the premature infant (Sturman *et al.*, 1970).

A single large dose of amino acid usually brings about a striking increase in the plasma concentration of that amino acid. After a single load of leucine, the level increased fifteenfold in the plasma of infants

(Snyderman *et al.*, 1968); when the same dose was given daily for
1 week, plasma leucine concentrations were normal. On the other hand,
infants did not adapt to loading with methionine, since plasma methionine concentration remained markedly elevated after 1 week.

Sauberlich (1961) measured PAA levels in young rats fed 6 percent
casein diets that contained an excessive amount of a given amino acid.
After 1 month on these diets, a high plasma concentration of the test
amino acid was usually found. Individual amino acids exhibited varying
degrees of toxicity. Glutamic acid was exceptional in that it had little
or no inhibitory effect on growth and that the plasma concentration of
the amino acid was only slightly elevated. This finding is hardly surprising, as Neame and Wiseman (1957) reported that glutamic acid is
transaminated in the intestinal wall and gives rise to increased alanine
levels in mesenteric blood, alanine being removed by the liver. In
Sauberlich's study, when the diet contained 5 percent of L-alanine,
plasma alanine concentration increased sixfold but growth remained
unaffected. All other amino acids tested individually at the 5 percent
level in the diet inhibited growth to some extent. The more toxic amino
acids—methionine, tryptophan, histidine, aspartic acid, and tyrosine—
accumulated excessively in the plasma. However, threonine had only a
slight effect on growth, even though plasma threonine concentrations
were also very high. Various aspects of amino acid toxicity, including
PAA concentrations and amino acid metabolism, have been reviewed
(Harper *et al.*, 1970) and are further discussed by A. E. Harper (pp.
138–166) in this volume.

RELATIONSHIP BETWEEN PLASMA AND TISSUE CONCENTRATIONS OF AMINO ACIDS

It has been reported that free amino acids of muscle, like those of
plasma, tend to reflect dietary excesses and deficiencies of amino acids
(Tannous *et al.*, 1966; Pawlak *et al.*, 1968). In one study (McLaughlan,
unpublished), plasma and muscle concentrations of free lysine were
measured at 3-h intervals in rats consuming a lysine-deficient diet. Initially, both plasma and muscle concentrations were low; but, when the
diet was removed and animals entered the fasting state, plasma and
muscle lysine concentrations increased at the same rate. After 16 h,
both reached levels characteristic of the normal fasting state. Pawlak
et al. (1968) reported that the difference in free lysine content of
muscle in rats fed diets containing low or high levels of lysine was much

greater than was the corresponding difference in plasma lysine concentrations.

Because the plasma pool is only a small fraction of the total tissue pool of free amino acids, comparison of relative amino acid patterns in some of the major pools is of interest. Table 1 shows amino acid concentrations in plasma, liver, muscle, and brain. In general, most amino acids are present at higher concentrations in liver and muscle than in either plasma or brain tissue; the exception is tryptophan, which is highest in plasma. All concentrations, except for threonine, which is much higher in the brain than in plasma, are lowest in brain. Although the data are not given, glutamic acid is normally 50–100 times higher in brain than in plasma. Clark *et al.* (1966) have shown that most of the dispensable amino acids are also present at remarkably high concentrations in liver and muscle.

In his review on free amino acid pools, Munro (1970) compared the proportion of each amino acid normally present in the various tissues of the body. The free amino acid content of the plasma pool is only about 2–3 percent of the total free amino acid pool, and the amino acid content of the total pool is only a small fraction of the daily requirement for amino acids. Christensen (1964) has pointed out, however, that the free amino acid pools of the tissues and plasma must serve as principal factors in keeping amino acid nutrition at the cellular level from being entirely spasmodic.

TABLE 1   Free Amino Acid Content of Various Rat Tissues[a]

Amino Acid	Tissue			
	Plasma	Muscle	Liver	Brain
Leu	24	33	51	15
Phe	7.7	15	21	5.6
Try	6.5	3.7	5.6	2.3
Val	24	35	38	15
His	7.4	67	56	11
Lys	31	36	40	19
Ile	15	23	21	4.8
Pro	18	38	49	12
Tyr	9.4	20	25	9.5
Met	6.6	17	17	8.1
Thr	40	97	136	111

[a]Expressed in μmol/100 ml of plasma or 100 g of tissue.
Recalculated from data of Williams *et al.* (1950).

PAA METHODS FOR EVALUATING PROTEINS

Many studies have dealt with the possible use of PAA concentrations
in the evaluation of proteins, but few of them were carried far enough
to be practical. Frame (1958) measured concentrations in adult humans
before, and at hourly intervals after, they consumed a large meal of
eggs. Most of the amino acids increased in concentration after the meal,
but the increases did not directly reflect the relative amino acid composition of the food itself.

The first truly comprehensive study designed to evaluate the feasibility of using PAA concentrations in assessing protein nutrition, however, was carried out by Longenecker and Hause (1959). In a first series
of experiments, in which diets contained 16 percent protein, PAA
changes in adult dogs were irregular; in some cases the values dropped
after the meal. In subsequent tests using diets containing 32 percent
protein, plasma concentrations of most amino acids rose. When the
relative changes in PAA concentrations after the meal were compared
to the amino acid composition of the ingested protein, no consistent
relationship was found.

The investigators then calculated the average increase, or decrease, of
each amino acid in the plasma and divided by the relative requirement
of the dog for each; this value, when multiplied by 100, gave what they
called the plasma amino acid (PAA) ratio. They applied the method to
wheat gluten, gelatin, and casein. The ranking of amino acids, according
to the extent of their relative deficiencies as indicated by PAA ratio,
paralleled closely the order found by chemical score. In an extension of
these studies, Longenecker and Hause (1961) fed two human adults test
meals consisting of wheat gluten, sugar, and corn oil slurried in orange
juice. The PAA ratio method presumably was valid in indicating lysine
to be the limiting amino acid in wheat gluten for humans.

One objection to the PAA ratio method was the large amount of protein that had to be ingested at a single meal (subjects consumed 85 g of
wheat gluten). Many foods, and almost all diets, contain less than 20
percent protein; and ingestion of the amount consumed in the investigations of Longenecker and Hause would require some form of protein
concentration. However, other studies of adult subjects indicated that
meaningful PAA ratios might be obtained after test meals containing as
little as 20–30 g of protein (McLaughlan *et al.*, 1963; Yearick and
Nadeau, 1967).

In studies with chicks, Hill and Olsen (1963) avoided the use of a
single high-protein meal by feeding amino acid-deficient test diets for
1 week before taking blood samples. They tested several diets and de-

termined modified PAA ratios. The method correctly predicted the
limiting amino acids as demonstrated by growth tests that involved
amino acid supplementation of diets. The principal feature of the PAA
ratio method is the use of the amino acid requirement pattern of the
animal as a means of assessing the significance of differences in PAA
patterns before and after test meals. This attribute could limit the value
of the method in clinical nutrition, due to the limited reliability of
values for human amino acid requirements.

Another method, called the PAA score, was developed from measurements on young rats fed test diets for several days (McLaughlan, 1964).
This procedure utilized the "fasting PAA pattern" (plasma level of each
essential amino acid after animals fasted 12–16 h) as a reference pattern.
PAA score was defined as the concentration of a given amino acid in
the plasma of the test group over the concentration of that amino acid
in the plasma of fasted rats, expressed as a percentage; the amino acid
with the lowest score was considered to be the limiting amino acid.
Although the choice of the fasting PAA pattern for reference purposes
was an arbitrary one, its use was advocated until the development of a
more suitable yardstick.

The PAA score has been applied to more than 40 individual foods
and mixed diets (McLaughlan, 1964; McLaughlan *et al.*, 1966, 1967).
In almost all cases in which the limiting amino acid was known, the
PAA score approach correctly identified it. However, a difficulty arose
in interpreting PAA data for methionine and cystine. PAA scores
showed (correctly) that the latter was the limiting amino acid in casein
for growth of rats (Bergen, 1968), but the high score for the former indicated that it also was adequate. In fact, either methionine or cystine
may be used to correct the cystine deficiency in casein.

As noted above, the fasting amino acid pattern of the chick is unsuit-
able as a reference; another approach involves the use of the PAA pattern of chicks fed an "ideal" diet. Dean and Scott (1962) developed a
crystalline amino acid diet containing what they believed to be an optimal level of each amino acid for growth of young chicks. Smith and
Scott (1965) used the PAA pattern of chicks fed the standard amino
acid diet as a reference pattern for scoring purposes. When the plasma
patterns from chicks fed the test proteins were compared with those fed
the reference diet, acute amino acid deficiencies in those fed the test
diets were readily apparent. In addition to a low plasma level of the
limiting amino acid, they exhibited apparent deficiencies of other amino
acids that, in fact, were adequate in the test proteins. Subsequent tests
confirmed that the crystalline amino acid diet contained slight excesses
of these amino acids, which circumstance accounted for the erroneous

results in the earlier PAA tests. These studies clearly demonstrated that the value of this PAA method for predicting adequacy of intact proteins for chick growth depended upon the extent to which the amino acid content of the standard reference diet coincided with the exact amino acid requirements of the chick. A similar conclusion was reached in studies with rats (McLaughlan, 1964), in which the "ideal" protein was whole egg.

In order to establish a "steady state" with respect to amino acid uptake into the bloodstream, Smith and Scott (1965) administered 800 mg of feed to chicks every 30 min for 6 h before sampling blood. They observed, however, that some diets tended to accumulate in the crop and that different diets did not reach the absorptive areas of the intestinal tract at equal rates.

ESTIMATION OF AMINO ACID REQUIREMENTS

When diets that contain widely different amounts of lysine are fed to young rats for 1 or 2 days, there is an almost direct relationship between plasma and dietary lysine concentrations (McLaughlan et al., 1961). However, if lysine is the limiting factor in diets that are fed for longer periods to young animals (rats, chicks, or pigs), the curve relating plasma lysine concentrations to that of the diet is either markedly sigmoid (Morrison et al., 1961; Zimmerman and Scott, 1965; Mitchell et al., 1968; Stockland et al., 1970) or else remains almost horizontal until intake approaches the requirement value and then deflects sharply upwards. In lysine studies, plasma threonine and lysine concentrations usually show a reciprocal relationship (Morrison et al., 1961).

Zimmerman and Scott (1965) used the "break" (i.e., the point where the curve bends upwards) in the plasma response curve as the end point in titrating the requirements of the chick for lysine, arginine, and valine; the predicted requirements agreed with those estimated by methods based on growth studies. This approach has also been used to estimate the lysine, isoleucine, leucine, and histidine requirements of the young pig (Mitchell et al., 1968) and the lysine needs of the weanling rat (Stockland et al., 1970). More recently, Young et al. (1971) assessed the value of this method for estimating the tryptophan requirement of human adults who consumed test diets for 4–6 days.

The postprandial plasma tryptophan response curves resembled plasma lysine response curves for animals fed graded levels of lysine (Morrison et al., 1961; Zimmerman and Scott, 1965; Mitchell et al., 1968; Stockland et al., 1970). Nitrogen balance studies carried out in the same subjects indicated a tryptophan requirement of 2–2.6 mg/kg body weight, whereas the break in the plasma tryptophan response curve occurred at about 3 mg/kg body weight. Young and co-workers have extended these investigations and have laid the ground work for similar studies with all the essential amino acids (Özalp *et al.*, 1972; Young *et al.*, 1972).

In studies with young rats (Morrison *et al.*, 1961; Stockland *et al.*, 1970), the break in the plasma lysine response curve occurred at an intake below the requirement indicated by the growth curve. The lysine content of the plasma at the point where the curve deflects upwards is very low (Morrison *et al.*, 1961; Zimmerman and Scott, 1965; Mitchell *et al.*, 1968; Stockland *et al.*, 1970; Young *et al.*, 1971); this is the prime characteristic of any limiting amino acid. Consequently, the requirement may be higher than indicated by the break on the plasma response curve.

McLaughlan and Illman (1967) fed test diets for only 3 days to young rats and obtained reasonably straight dose–response curves for lysine, tryptophan, and leucine, whereas curves for threonine, tryptophan, and isoleucine tended to be sigmoid. They reported that the point at which the response curve reached the normal fasting value (PAA score of 100 for test amino acid) was a suitable end point in titrating amino acid requirements of growing rats. Estimated requirements for the six amino acids tested agreed satisfactorily with average values obtained by other methods as reported recently in the literature.

The PAA method might be particularly useful for titrating the correct level of amino acid supplementation of diets (special case of amino acid requirements). When the question of the most appropriate end point in the titration of amino acid requirements is resolved, it would be appropriate to apply this approach to the problem of lysine supplementation of cereal diets for children.

PAA CONCENTRATIONS AND DISEASE

Screening for abnormalities in neonatal amino acid patterns is now widely practiced (Holt and Snyderman, 1964; Scriver, 1969). A common feature of such diseases (phenylketonuria, tyrosinemia, histidinemia, etc.) is a high serum concentration of the particular amino acid that is being metabolized abnormally. Phenylalanine concentration is one of the criteria used in clinical diagnosis of phenylketonuria. Consumption of a low-phenylalanine diet early in life is believed to result in normal development and intelligence; in such instances, plasma phenylalanine is monitored to assess adequacy of the diet (Holt and Snyderman, 1964). The transient hypermethioninemia reported in infants fed high levels of milk protein (Komrower and Robins, 1969) requires further study.

Liver damage induced by carbon tetrachloride or by diet may produce marked transient alterations in PAA patterns (Emery and Beveridge, 1961; Truhaut et al., 1966). Low plasma–arginine and high plasma–ornithine concentrations appear to result from release of arginase from damaged hepatic cells (Emery and Beveridge, 1961). PAA patterns of chronic alcoholics also show significant deviations from the normal (Siegel et al., 1963).

PAA concentrations are drastically altered in those suffering severe protein malnutrition. Studies have shown that PAN (free plasma–amino nitrogen) values in humans correlate only fairly well with protein nutrition (Albanese, 1959). Measurement of individual amino acids in plasma revealed that the concentrations of essential amino acids usually were low in instances of protein deficiency, but that plasma concentrations of the dispensable amino acids were often elevated (Holt et al., 1963); consequently, the overall change in PAN concentration was minimal. Arroyave (1962) examined changes in concentration of individual amino acids in plasma of children on a protein-free diet for several days. Concentrations of the essential amino acids and tyrosine fell rapidly for several days and then fluctuated daily around a low value. Concentrations of plasma alanine, glycine, and proline showed marked increases initially but fell to original values after 15 days. Arroyave concluded that measurement of one or several of the plasma amino acid concentrations might be more indicative of subclinical protein deficiencies than a measurement of PAN concentration.

Holt et al. (1963) presented plasma aminograms for 64 kwashiorkor patients from nine countries. Similar plasma amino acid patterns, which differed markedly from the normal, were found in patients from all countries studied. In general, the dispensable amino acids were present at normal or elevated concentrations; those of essential amino acids were low. Later, Holt and Snyderman (1964) published typical plasma aminograms of children with mild and with advanced protein deficiency. In serum of children with moderate protein deficiency, the concentrations of most amino acids were close to normal; those of alanine and glycine were high; and those of valine, leucine, and tyrosine were low. Practically all amino acids were present at reduced concentrations in serum of children with advanced protein deficiency, but glycine was present at near-normal concentrations; values for valine, leucine, and tyrosine were very low. Whitehead and Dean (1964) reported similar, but not identical, PAA patterns in relation to protein deficiency.

Whitehead (1964) described a rapid method, using one-dimensional paper chromatography, for detecting abnormal PAA patterns. The ratio of a group of dispensable amino acids (primarily glycine, serine, and taurine) to a group of essential amino acids (primarily leucine, isoleucine, and methionine) proved useful for detecting protein deficiency. McLaren *et al.* (1965) did not find the method feasible for identifying potential cases of protein malnutrition, but other workers (Poey *et al.*, 1967; Arroyave and Bowering, 1968) have since confirmed the potential of the plasma method for detecting the subclinical cases.

PAA patterns of children with kwashiorkor suggest that plasma concentrations of certain amino acids are better indicators of protein nutrition than are others. A complete aminogram permits the calculation of such potentially useful ratios as phenylalanine–tyrosine and valine–glycine (Arroyave *et al.*, 1962; Whitehead and Dean, 1964). Both of these ratios may be better indicators of protein malnutrition than is the ratio of essential to dispensable amino acids proposed by Whitehead (1964). However, Swendseid *et al.* (1966) reported that plasma–valine responds so rapidly to a low-protein intake that a low plasma-valine value is not a reliable indicator of a true protein depletion state, unless it occurs in conjunction with decreased concentrations of other essential amino acids. Weller *et al.* (1969) also found a low concentration of plasma valine in adult subjects fed a nonprotein diet for only 1 day and an only slightly lower level after 2 weeks. Inter- and intrasubject variation in PAA levels was wide, but the authors concluded that fasting PAA patterns were more characteristic of the diet than of short-term nutritional status. Although PAA methods are not fully satisfactory in diagnosis of protein malnutrition, such studies appear to be a valuable adjunct to such other methods as the determination of plasma proteins.

PLASMA PROTEINS IN PROTEIN MALNUTRITION

Although it has frequently been observed that there is a reduction in the amount of total plasma protein in children with kwashiorkor, this alteration has been of little value for diagnostic purposes (Arroyave, 1962; Krehl and Hodges, 1965). In subjects with protein–calorie malnutrition, plasma albumin concentration is usually low, but the total protein concentration is often close to normal; the latter results from elevated globulin concentrations commonly associated with infection. However, the possibility of changes in plasma volume complicates the interpretation of alterations in plasma protein (Albanese, 1959; Arroyave, 1962). Nevertheless, serum protein measurements, particularly of the albumin fraction, are widely used as indicators of protein status in nutrition surveys

(Kelsay, 1969) and during clinical treatment for kwashiorkor. Heard *et al.* (1969) reported that measurement of serum protein concentrations in puppies provided a better index of protein nutritional status than did the amino acid ratio (Whitehead, 1964).

In the rat (Kirsch *et al.*, 1968), the rate of albumin synthesis by the liver is regulated, in part at least, by the protein adequacy of the diet. In studies with perfused livers, low concentrations of valine, leucine, and isoleucine in the perfusate reduced the rate of albumin synthesis, whereas the addition of branched-chain amino acids to the perfusate restored the synthesis rate to normal (Kirsch *et al.*, 1969). Recent studies with chicks (Thomas and Combs, 1967) also showed that, although plasma albumin concentration related directly to the protein content of the diet, it also related inversely to caloric intake and weight gain. Consequently, chicks fed a low protein diet *ad libitum* had low plasma albumin levels, whereas chicks fed restricted amounts of the same diet had "normal" plasma albumin levels. These studies help clarify the reason for hypoalbuminemia in children with kwashiorkor (primarily a protein deficiency) as opposed to the near-normal albumin associated with marasmus (deficiency of both protein and calories).

Graham *et al.* (1966) reported that weight gain and increase in serum albumin concentration in young children did not necessarily correlate during treatment of kwashiorkor. When caloric intake was high, thus favoring rapid weight gain and nitrogen retention, serum albumin was adversely affected, even when the protein was cow's milk and the level of dietary protein was apparently adequate. The value of serum albumin under field conditions is not well established. Rutishauser and Whitehead (1969) studied protein–calorie malnutrition in three areas of Uganda, where children were living under different environmental and dietary conditions. Measuring a variety of diagnostic variables for detecting malnutrition, they found reduced serum albumin and altered amino acid ratios; but the correlation between these criteria, or between either of these criteria and other criteria of protein malnutrition, was not high. The writers pointed out that infestations of hookworm and other parasites affect certain variables much more than others. They concluded that many difficulties must be resolved before biochemical tests for mild protein deficiency can be used with confidence—a situation equally true of clinical examinations and anthropometric measurements.

SUMMARY

Published information on PAA concentrations is extensive, repetitious, and frequently controversial. Although the concentrations usually reflect the amino acid composition of dietary proteins, numerous other factors play important roles. Early enthusiasm for a rapid *in vivo* method for rating protein quality now appears to have been premature. Nevertheless, a few limited, but potentially valuable, uses for PAA concentrations seem assured. Although the plasma method does not necessarily give a reliable indication of the extent of the amino acid deficiency, several groups of workers have demonstrated that measurement of PAA concentrations can provide a guide to identifying the dietary amino acid that is limiting. In addition, evidence indicates that plasma concentrations may provide a rapid method for estimating amino acid requirements of young experimental animals; presumably this approach would be fruitful in studies of humans. The PAA pattern of children with kwashiorkor is grossly abnormal, a finding that has potential value in the clinical diagnosis of various stages of protein deficiency.

Prolonged intake of a protein-deficient diet without caloric restriction results in low serum albumin; regeneration of plasma albumin is now an established criterion of adequate treatment of kwashiorkor. However, much work needs to be done to clarify the value of changes in serum protein concentrations in the detection of early or mild cases of protein malnutrition.

REFERENCES

Albanese, A. A. 1959. Criteria of protein nutrition. Pages 297–347 *in* A. A. Albanese, ed. Protein and amino acid nutrition. Academic Press, New York.

Anderson, H. L., and H. Linkswiler. 1969. Effect of source of dietary nitrogen on plasma concentration and urinary excretion of amino acids of men. J. Nutr. 99: 91–100.

Anderson, H. L., N. J. Benevenga, and A. E. Harper. 1968. Associations among food and protein intake, serine dehydratase, and plasma amino acids. Am. J. Physiol. 214:1008–1013.

Arroyave, G. 1962. The estimation of relative nutrient intake and nutritional status by biochemical methods: Proteins. Am. J. Clin. Nutr. 11:447–461.

Arroyave, G., and J. Bowering. 1968. Plasma free amino acids as an index of protein nutrition. Evaluation of Whitehead's method. Arch. Latinoam. Nutr. 38: 341–361.

Arroyave, G., D. Wilson, C. de Funes, and M. Béhar. 1962. The free amino acids in blood plasma of children with kwashiorkor and marasmus. Am. J. Clin. Nutr. 11:517–524.

Bergen, W. G., D. B. Purser, and J. H. Cline. 1968. Determination of limiting amino acids of rumen-isolated microbial proteins fed to rat. J. Dairy Sci. 51:1698–1700.

Boomgaardt, J., and B. E. McDonald. 1969. Comparison of fasting plasma amino acid patterns in the pig, rat, and chicken. Can. J. Physiol. Pharmacol. 47:392–395.

Bromer, W. W., and R. E. Chance. 1969. Zinc glucagon depression of blood amino acids in rabbits. Diabetes 18:748–754.

Christensen, H. N. 1964. Free amino acids and peptides in tissues. Pages 105–124 *in* H. N. Munro and J. B. Allison, eds. Mammalian protein metabolism. Vol. I. Academic Press, New York.

Clark, A. J., Y. Peng, and M. E. Swendseid. 1966. Effect of different essential amino acid deficiencies on amino acid pools in rats. J. Nutr. 90:228–234.

Dean, W. F., Jr., and H. M. Scott. 1962. The development of an amino acid standard for the early growth of chicks. Poult. Sci. 41:1640. (A)

Denton, A. E., and C. A. Elvehjem. 1954. Availability of amino acids *in vivo*. J. Biol. Chem. 206:449–460.

Emery, G. N., and J. M. R. Beveridge. 1961. The cause of the disappearance of arginine from the blood of rats with acute hepatic necrosis induced by dietary means. Can. J. Biochem. Physiol. 39:977–980.

Feigin, R. D., A. S. Klainer, and W. R. Biesel. 1968. Factors affecting circadian periodicity of blood amino acids in man. Metabolism (Clin. Exp.) 17:764–775.

Frame, E. G. 1958. The levels of individual free amino acids in the plasma of normal man at various intervals after a high-protein meal. J. Clin. Invest. 37:1710–1723.

Gitler, C. 1964. Protein digestion and absorption in nonruminants. Pages 35–69 *in* H. N. Munro and J. B. Allison, eds. Mammalian protein metabolism. Vol. I. Academic Press, New York.

Graham, G. G., A. Cordano, and J. M. Baertl. 1966. Studies in infantile malnutrition. IV. The effect of protein and calorie intake on serum proteins. Am. J. Clin. Nutr. 18:11–15.

Guggenheim, K., S. Halevy, and N. Friedmann. 1960. Levels of lysine and methionine in portal blood of rats following protein feeding. Arch. Biochem. Biophys. 91:6–10.

Harper, A. E., and Q. R. Rogers. 1965. Amino acid imbalance. Proc. Nutr. Soc. 24:173–190.

Harper, A. E., N. J. Benevenga, and R. M. Wohleuter. 1970. Effects of ingestion of disproportionate amounts of amino acids. Physiol. Rev. 50:428–558.

Heard, C. R. C., S. M. Kreigsman, and B. S. Platt. 1969. The interpretation of plasma amino acid ratios in protein–calorie deficiency. Br. J. Nutr. 23:203–210.

Henderson, L. M., P. E. Schurr, and C. A. Elvehjem. 1949. The influence of fasting and nitrogen deprivation on the concentration of free amino acids in rat plasma. J. Biol. Chem. 177:815–823.

Hill, D. C., and E. M. Olsen. 1963. Effect of starvation and a nonprotein diet on blood plasma amino acids, and observations on the detection of amino acids limiting growth of chicks fed purified diets. J. Nutr. 79:303–310.

Holt, L. E., Jr., and S. E. Snyderman. 1964. Anomalies of amino acid metabolism. Pages 321–372 *in* H. N. Munro and J. B. Allison, eds. Mammalian protein metabolism. Vol. 2. Academic Press, New York.

Holt, L. E., Jr., S. E. Snyderman, P. M. Norton, E. Roitman, and J. Finch. 1963. The plasma aminogram in kwashiorkor. Lancet 2:1343–1348.

Kelsay, J. L. 1969. A compendium of nutritional status studies and dietary evaluation studies conducted in the United States, 1957–1967. J. Nutr. 99:(Suppl. 1, Part II):123–166.

Kirsch, R. E., L. Frith, E. Black, and R. Hoffenberg. 1968. Regulation of albumin synthesis and catabolism by alteration of dietary protein. Nature 217:578–579.

Kirsch, R. E., S. J. Saunders, L. Frith, S. Wicht, L. Kelman, and J. F. Brock. 1969. Plasma amino acid concentration and the regulation of albumin synthesis. Am. J. Clin. Nutr. 22:1559–1562.

Knipfel, J. E., H. G. Botting, F. J. Noel, and J. M. McLaughlan. 1969. Amino acids in blood plasma and tissues of rats following glucose force-feeding. Can. J. Biochem. 47:323–327.

Komrower, G. M., and A. J. Robins. 1969. Plasma amino acid disturbance in infancy. I. Hypermethioninaemia and transient tyrosinaemia. Arch. Dis. Child. 44:418–421.

Krehl, W. A., and R. E. Hodges. 1965. The interpretation of nutrition survey data. Am. J. Clin. Nutr. 17:191–199.

Leung, P. M-B., and Q. R. Rogers. 1969. Food intake: Regulation by plasma amino acid pattern. Life Sci. 8(Part 2):1–9.

Longenecker, J. B., and N. L. Hause. 1959. Relationship between plasma amino acids and composition of the ingested protein. Arch. Biochem. Biophys. 84:46–59.

Longenecker, J. B., and N. L. Hause. 1961. Relationship between plasma amino acids and composition of the ingested protein. II. A shortened procedure to determine plasma amino acid (PAA) ratios. Am. J. Clin. Nutr. 9:356–362.

McLaren, D. S., W. W. Kamel, and N. Ayyoub. 1965. Plasma amino acids and the detection of protein–calorie malnutrition. Am. J. Clin. Nutr. 17:152–157.

McLaughlan, J. M. 1964. Blood amino acid studies. V. Determination of the limiting amino acid in diets. Can. J. Biochem. 42:1353–1360.

McLaughlan, J. M., and W. I. Illman. 1967. Use of free plasma amino acid levels for estimating amino acid requirements of the growing rat. J. Nutr. 93:21–24.

McLaughlan, J. M., F. J. Noel, A. B. Morrison, and J. A. Campbell. 1961. Blood amino acid studies. I. A micromethod for the estimation of free lysine, methionine, and threonine. Can. J. Biochem. Physiol. 39:1669–1674.

McLaughlan, J. M., F. J. Noel, A. B. Morrison, and J. A. Campbell. 1963. Blood amino acid studies. IV. Some factors affecting plasma amino acid levels in human subjects. Can. J. Biochem. Physiol. 41:191–199.

McLaughlan, J. M., F. J. Noel, and S. Venkat Rao. 1966. Prediction of limiting amino acids in dietary proteins. Pages 333–337 in Joachim Küehnau, ed. Proc. 7th Int. Congr. Nutr., International Union of Nutritional Sciences, Hamburg.

McLaughlan, J. M., S. Venkat Rao, F. J. Noel, and A. B. Morrison. 1967. Blood amino acid studies. VI. Use of plasma amino acid score for predicting limiting amino acid(s) in dietary proteins. Can. J. Biochem. 45:31–37.

Manchester, K. L. 1970. The control of insulin of amino acid accumulation in muscle. Biochem. J. 117:457–465.

Mitchell, J. R., D. E. Becker, A. H. Jensen, B. G. Harmon, and H. W. Norton. 1968. Determination of amino acid needs of the young pig by nitrogen balance and plasma-free amino acids. J. Anim. Sci. 27:1327–1331.

Morrison, A. B., E. J. Middleton, and J. M. McLaughlan. 1961. Blood amino acid studies. II. Effects of dietary lysine concentration, sex, and growth rate on plasma free lysine and threonine levels in the rat. Can. J. Biochem. Physiol. 39:1675–1680.

Munro, H. N. 1964. A general survey of pathological changes in protein metabolism.

Pages 267–319 *in* H. N. Munro and J. B. Allison, eds. Mammalian protein metabolism. Vol. 2. Academic Press, New York.

Munro, H. N. 1970. Free amino acid pools and their role in regulation. Pages 299–386 *in* H. N. Munro, ed. Mammalian protein metabolism. Vol. 4. Academic Press, New York.

Nasset, E. S. 1968. Contribution of the digestive system to the amino acid pool. Pages 80–87 *in* J. H. Leathem, ed. Protein nutrition and free amino acid patterns. Rutgers University Press, New Brunswick, N.J.

Neame, K. D., and G. Wiseman. 1957. The transamination of glutamic and aspartic acids during absorption by the small intestine of the dog *in vivo*. J. Physiol. 135: 442–450.

Özalp, I., V. R. Young, J. Nagchaudhuri, K. Tontisirin, and N. S. Scrimshaw. 1972. Plasma amino acid response in young men given diets devoid of single essential amino acids. J. Nutr. 102:1147–1158.

Pawlak, M., R. Pion, M. Allez, C. Boyle, and J. Leroux. 1968. Influence de la supplémentation des protéines de blé par des doses croissantes de lysine sur la teneur en acides aminés libre du sang et du muscle du rat en croissance. Ann. Biol. Anim. Biochim. Biophys. 8:517–530.

Poey, S. H., C. S. Rose, Muhilal, and S. Zuraida. 1967. Serum free amino acids in children with protein–calorie deficiency. Am. J. Clin. Nutr. 20:1295–1299.

Potter, E. L., D. B. Purser, and J. H. Cline. 1968. Effect of various energy sources upon plasma free amino acids in sheep. J. Nutr. 95:655–663.

Rogers, Q. R., M-L. Chen, C. Peraino, and A. E. Harper. 1960. Observations on protein digestion in vivo. III. Recovery of nitrogen from the stomach and small intestine at intervals after feeding diets containing different proteins. J. Nutr. 72:331–339.

Rutishauser, I. H. E., and R. G. Whitehead. 1969. Field evaluation of two biochemical tests which may reflect nutritional status in three areas of Uganda. Br. J. Nutr. 23:1–13.

Sauberlich, H. E. 1961. Studies on the toxicity and antagonism of amino acids for weanling rats. J. Nutr. 75:61–72.

Scriver, C. R. 1969. Inborn errors of amino-acid metabolism. Br. Med. Bull. 25: 35–41.

Siegel, F. L., M. K. Roach, and W. B. Deville. 1963. Plasma amino acid patterns in alcoholics: Ethanol induced modification. Fed. Proc. 22:680. (A)

Smith, R. E., and H. M. Scott. 1965. Use of free amino acid concentrations in blood plasma in evaluating the amino acid adequacy of intact proteins for chick growth. I. Free amino acid patterns of blood plasma of chicks fed unheated and heated fishmeal proteins. J. Nutr. 86:37–44.

Snyderman, S. E., L. E. Holt, Jr., P. M. Norton, and E. Roitman. 1968. Effect of high and low intakes of individual amino acids on the plasma aminogram. Pages 19–31 *in* J. H. Leathem, ed. Protein nutrition and free amino acid patterns. Rutgers University Press, New Brunswick, N.J.

Snyderman, S. E., L. E. Holt, Jr., P. M. Norton, and S. V. Phansalkar. 1970. Protein requirement of the premature infant. II. Influence of protein intake on free amino acid content of plasma and red blood cells. Am. J. Clin. Nutr. 23:890–895.

Stockland, W. L., R. J. Meade, and A. L. Melliere. 1970. Lysine requirement of the growing rat: Plasma-free lysine as a response criterion. J. Nutr. 100:925–933.

Sturman, J. A., G. Gaull, and N. C. R. Raiha. 1970. Absence of cystathionase in human fetal liver: Is cystine essential? Science 169:74–76.

Swendseid, M. E., S. G. Tuttle, W. S. Figueroa, D. Mulcare, A. J. Clark, and F. J. Massey. 1966. Plasma amino acid levels of men fed diets differing in protein content. Some observations with valine-deficient diets. J. Nutr. 88:239–248.

Swendseid, M. E., S. G. Tuttle, E. J. Drenick, C. B. Joven, and F. J. Massey. 1967. Plasma amino acid response to glucose administration in various nutritive states. Am. J. Clin. Nutr. 20:243–249.

Swendseid, M. E., C. Y. Umezawa, and E. Drenick. 1969. Plasma amino acid levels in obese subjects before, during, and after starvation. Am. J. Clin. Nutr. 22:740–743.

Tannous, R. I., Q. R. Rogers, and A. E. Harper. 1966. Effect of leucine–isoleucine antagonism on the amino acid pattern of plasma and tissues of the rat. Arch. Biochem. Biophys. 113:356–361.

Thomas, O. P., and G. F. Combs. 1967. Relationship between serum protein level and body composition in the chick. J. Nutr. 91:468–472.

Truhaut, R., J. C. Delarue, and C. Bohuon. 1966. Sur l'entérêt comme test d'imprégnation et de pronostic de l'étude des variations de l'amino-acidémie de l'intoxication par le tétrachlorure de carbone. Etudes expérimentales chez le rat. Ann. Biol. Clin. (Paris) 24:727–738.

Weller, L. A., S. Margen, and D. H. Calloway. 1969. Variation in fasting and post-prandial amino acids of men fed adequate or protein-free diets. Am. J. Clin. Nutr. 22:1577–1583.

Whitehead, R. G. 1964. Rapid determination of some plasma amino acids in sub-clinical kwashiorkor. Lancet 1:250–252.

Whitehead, R. G., and R. F. A. Dean. 1964. Serum amino acids in kwashiorkor. I. Relationship to clinical condition. Am. J. Clin. Nutr. 14:313–319.

Williams, J. N., Jr., P. E. Schurr, and C. A. Elvehjem. 1950. The influence of chilling and exercise on free amino acid concentrations in rat tissues. J. Biol. Chem. 182:55–59.

Wurtman, R. J. 1970. Diurnal rhythms in mammalian protein metabolism. Pages 445–479 in H. N. Munro, ed. Mammalian protein metabolism. Vol. 4. Academic Press, New York.

Wynne, E. S., and C. L. Cott. 1956. Effect of food intake on amino acids in human plasma. Am. J. Clin. Nutr. 4:275–278.

Yearick, E. S., and R. G. Nadeau. 1967. Serum amino acid response to isocaloric test meals. Am. J. Clin. Nutr. 20:338–344.

Young, V. R., and J. Zamora. 1968. Effects of altering the proportions of essential to nonessential amino acids on growth and plasma amino acid levels in the rat. J. Nutr. 96:21–27.

Young, V. R., M. A. Hussein, E. Murray, and N. S. Scrimshaw. 1971. Plasma trypto-phan response curve and its relation to tryptophan requirements in young adult men. J. Nutr. 101:45–59.

Young, V. R., K. Tontisirin, I. Özalp, F. Lakshmamam, and N. S. Scrimshaw. 1972. Plasma amino acid response curve and amino acid requirements in young men: Valine and lysine. J. Nutr. 102:1159–1170.

Zimmerman, R. A., and H. M. Scott. 1965. Interrelationship of plasma amino acid levels and weight gain in the chick as influenced by suboptimal and superoptimal dietary concentrations of single amino acids. J. Nutr. 87:13–18.

G. G. GRAHAM

Effects of Deficiency of
Protein and Amino Acids

To formulate a satisfactory definition of protein or amino acid
"deficiency" is difficult. Deficiency in an organism may be considered
to commence at a point when the depletion of protein and amino acids
no longer permits that organism to cope with a stressful situation without compromising function. Such a definition implies "stores" or "reserves" of "labile" protein that can be lost from the body and totally
regenerated without compromise of function or without permanent
damage (Munro, 1964; Ashley and Fisher, 1967). If deficiency is defined by the first of these criteria—i.e., compromise of function—such
"stores" are almost certainly less than 5 percent of body protein. If it is
defined by the second criterion, that of permanent damage, the "stores"
would be considerably greater; because very large amounts of muscle
protein can be lost from the body, causing a recognizable compromise
of function and marked alteration of body composition (Graham *et al.*,
1969a; Cheek *et al.*, 1970), but with the potential for complete regeneration being retained.

Researchers have expended much effort in trying to duplicate in
laboratory animals the complex syndrome of human malnutrition and
have paid considerable attention to the relations among energy intake,
total nitrogen intake, supply of individual amino acids, and the balance
among these. Nevertheless, controversies continue over the relative

109

importance of energy and protein and over the role of specific amino acid deficiencies, if any, in the practical problem of malnutrition. Much of the difficulty stems from poor understanding of important species differences, of human adaptability, and of the precise role of infection, the growth process, and the hormonal and enzymatic mechanisms that regulate the utilization of nutrients.

National and regional food balance tables, as well as nutrition surveys, fail to bring out individual variations in intake. When availability of food is reported as so many calories or so much protein per capita, exact determination of the foods given or denied each member of the family is impossible. This is particularly true of the vulnerable infant or preschool child, because of cultural or physiological considerations or because of the availability of time and conveniences to the mother, or her substitute, who must prepare and dispense food. Prevalence studies are handicapped by the lack of simple measures of nutritional status, particularly in regard to protein, that might identify the several cases of marginal deficiency that almost certainly exist for each case of overt clinical malnutrition.

DEFICIENCY IN MAN

Increases in the adult size of the Japanese, initially in the United States, and more recently in their own country, are striking proof of the ability of a whole race of human beings successfully to adapt to a low available food supply through failure to attain potential body size (Kimura, 1967). Only when food supply is abundant is genetic endowment the primary determinant of the size of the individual (Dugdale *et al.*, 1970); familial growth patterns are masked by adverse environmental influences, particularly limited availability and quality of food (Graham and Adrianzen, 1970). Even when seemingly adequate food supplies are available in the home, impediments such as those noted above may limit the food intake of infants and small children, particularly during episodes of infection. Inefficient utilization of ingested nutrients occurs not only as a result of obvious febrile infections but also of the more insidious and not so apparent ones, such as primary tuberculosis and pyelonephritis. Intestinal losses due to malabsorption or diarrhea are of major significance.

Interracial anthropometric comparisons must take nutrition into account. The ability of many groups to exist and multiply, despite intakes of protein that would rapidly lead to acute deficiency states in other groups, is attributable to their proportionately smaller size, lean body mass, and total energy intake, and to efficient adaptation. Little evidence exists of adverse effects from the smaller size, per se; indeed, much experimental evidence indicates that it is advantageous (Ross, 1964).

Prenatal Nutrition

Distinguishing the effects of nutrition from those of other environmental factors in evaluating the outcome of pregnancy has always been difficult. The Food and Nutrition Board (FNB, 1970) has published a review of the effects of maternal nutrition on the outcome of pregnancy, including an extensive bibliography. Studies of mass starvation during World War II and in chronically deprived populations suggest that severe deficits of calories and protein result in decreased fertility, in a decrease in the length and weight of the newborn, and in increased rates of neonatal mortality. The evidence for an adverse effect of moderate lifelong, prepregnancy, or gestational dietary deficiency of calories or protein is more meager. It seems likely that if the mother enters pregnancy in a satisfactory nutritional state, or receives an adequate diet during pregnancy, particularly during the last trimester, her offspring at birth will be little affected, at least in size and weight. Considerable additional research is needed to clarify this point.

Infancy and Childhood

To sort out and understand the effects of protein or amino acid deficiencies in growing human beings without understanding the cellular events and the regulatory mechanisms that constitute normal growth is impossible. Despite significant advances (Cheek, 1968), wide gaps in our knowledge still exist. Early growth is the result of a combination of cellular multiplication and hypertrophy, with different schedules for different tissues. The stimulatory effects of maternal and placental hormones probably carry over into early extrauterine life, until the infant's own mechanisms take over. Thyroid hormone influences cellular multiplication, particularly of the supporting tissues, during the entire growth period, with growth hormone and eventually the sex hormones assuming important roles. Insulin and growth hormone are vital to growth of cells, particularly muscle cells. Adrenal glucocorticoids are antagonistic to cell multiplication and to muscle cell hypertrophy (Munro, 1964). Effective energy intake is the dietary factor of greatest importance in stimulating cell multiplication. An adequate protein intake is vital to cell growth, and the amino acid tryptophan plays a singular role in ribosomal aggregation during protein synthesis (Fleck `et al.`, 1965). The utilization of amino acids, of dietary or endogenous origin, is limited by the supply of the one, or ones, present in least amount relative to requirements for protein synthesis at a given site. Infections curtail growth by affecting intake, absorption, utilization, and expenditure of nutrients.

Hypocaloric dwarfism   The effects of nutrient deficiencies on growth depend on the degree of severity, duration, and timing relative to growth phases. Millions of children who are deprived or who suffer from diseases that interfere with nutrient intake and utilization experience, as a normal course of events, restriction of both energy and protein intake from the second semester of life until sexual maturation. The end result is significant stunting. Although some studies suggest that permanent deficits in stature, musculature, and brain size result from finite periods of malnutrition (Graham, 1967; Stoch and Smythe, 1967; Winick and Rosso, 1969) and that "catch-up" growth is limited in duration (Prader *et al.*, 1963) others suggest that the principal effect of malnutrition is a maturational lag and that "catch-up" potential may be limited only by the time remaining before the achievement of full sexual maturity (Garrow and Pike, 1967; Graham, 1968). The healthy, fullterm newborn is able to survive semistarvation during most of the first 4 months and yet make up, completely and promptly, extremely severe deficits (Chase and Martin, 1970).

Moderate chronic or intermittent energy deficits that are sufficient to arrest growth in most tissues have the effect of decreasing efficiency of protein and amino acid utilization. Dietary protein in amounts over and above that needed for maintenance is used as a source of energy. Total body composition corresponds to that of the "biologic" age, which is best estimated from body length, while cellular composition usually is normal (Cheek, 1968). Unduly prolonged breast feeding; excessive dilution of milk, intentionally or inadvertently; insufficiently frequent feeding; and steatorrhea from various causes can result in a state of "hypocaloric dwarfism."

Marasmus   If energy deficit is so severe as not to meet basal requirements for prolonged periods, activity decreases and maintenance protein needs are not met in most tissues, regardless of the protein intake. Important and poorly understood adaptations are then necessary for survival, and the clinical picture of marasmus develops. Arrested cellular multiplication occurs, compounded by atrophy of many tissues, with decreases in cell size (Frenk *et al.*, 1957; Garn *et al.*, 1964; Brunser *et al.*, 1966; Bradfield *et al.*, 1969; Cheek *et al.*, 1970) and curtailment of exocrine secretion (Barbezat and Hansen, 1968). The cellular integrity and function of some organs, most notably the liver, are conserved; enzyme activity is altered to favor survival and the most efficient utilization of the limited supply of exogenous and endogenous nutrients (Waterlow, 1968). Markedly reduced insulin secretion, persisting well into the period of rehabilitation, seems to be an integral part of adaptation (Graham *et al.*, 1969a; James and Coore, 1970). The status of adrenal, thyroid, and growth hormone secretion is unclear. For the latter, high fasting levels and blunted responses seem to be the rule (Graham *et al.*, 1969a). Although energy deficiency is the overriding determinant, dietary protein intake is almost invariably deficient as well; and protein metabolism is so severely affected that marasmus is usually included with kwashiorkor and intermediate forms under the broad term protein–calorie malnutrition, or PCM (Jelliffe *et al.*, 1954). Marasmus is most commonly the result of very early weaning to grossly inadequate liquid concoctions, or to the repeated and prolonged "therapeutic" starvation that is practiced in the treatment of repeated diarrheal episodes.

Kwashiorkor   The dramatic clinical picture of kwashiorkor (edema, hypoalbuminemia, and a fatty liver, with or without skin and hair changes) represents acute decompensation of a relatively long-standing deficiency state, usually precipitated by infection (Jelliffe *et al.*, 1954). In most cases, it is characterized by a gradual depletion of intra- and extracellular protein and amino acids (Viteri *et al.*, 1964), most strikingly in the liver (Waterlow, 1968), pancreas (Barbezat and Hansen, 1968), intestinal mucosa (Brunser *et al.*, 1966), hair (Bradfield *et al.*, 1968), and muscle (Frenk *et al.*, 1957). Simultaneously, a marked loss of those ions that are intimately linked to the metabolism of protein in cells—potassium (Alleyne *et al.*, 1970), magnesium (Montgomery, 1961), and phosphate (Waterlow and Wills, 1960)—occurs. Of particular importance to the clinical picture of acute protein deficiency are losses of liver protein, particularly enzymes (Waterlow, 1968), and the accumulation of fat and glycogen in hepatocytes. The fat, derived from both dietary and endogenous sources, probably accumulates as a result of inadequate synthesis of carrier β-lipoproteins (Truswell *et al.*, 1969). Other serum proteins, notably albumin (James and Hay, 1968) and ceruloplasmin (Graham and Cordano, 1969), are synthesized in inadequate amounts. The insufficient dietary supply of essential amino acids and of total nitrogen results in a characteristic decrease of the plasma and intracellular levels of most of the amino acids and an elevation of the dispensable ones (Holt *et al.*, 1963). Kwashiorkor is most commonly precipitated by an infection, such as measles, in a child who has been weaned to a diet of unsupplemented cereals or tubers.

The development of simple techniques for the identification of those children who are protein-depleted or marginally deficient represents an important challenge (IUNS, 1970). In this situation, growth has usually slowed down, permitting the use of simple anthropometric techniques. These, however, are not entirely satisfactory. Low values for height-for-age or weight-for-age can be of constitutional origin or the result of previous episodes of disease. Low weight-for-height values, which are somewhat more useful in diagnosis, can also be of constitutional origin, but more commonly are the result of energy and/or protein deficiency or of chronic infection. Human infants or children, as distinct from the rat, may consume for many weeks a diet adequate or nearly adequate in energy but deficient in protein or one of the essential amino acids. If they do, without converting the relative excess of energy to heat (Ashworth, 1969), continued cellular multiplication and growth and a state of acute protein deficiency may develop in a child of normal height and weight, typified by the "sugar baby" (Jelliffe *et al.*, 1954). More commonly, however, children on a protein-deficient diet gradually decrease total food intake, stop growing, or continue to do so at a much decelerated rate. Body composition begins to change, with decreases in muscle and visceral cell size. The creatinine excretion/height ratio decreases, making this one of the potentially useful tools for detecting this state (Viteri and Alvarado, 1970). Total body potassium and intracellular water, which are much less easily measured, are almost certainly decreased as well, whereas extracellular water increases. Mid-arm circumference is a fairly reliable predictor of muscle mass (Standard *et al.*, 1959). Serum albumin and the plasma amino acids are potentially useful indicators of protein status, the latter being overly sensitive to recent diet.

Specific amino acid deficiencies   If total intake of a dietary protein of poor biological value is not adequate to meet the requirements of one of the essential amino acids, one might expect to encounter manifestations of specific amino acid deficiency. In common dietary practice, the ones most likely to be limiting are lysine, methionine, and tryptophan. Although alterations in serum or plasma concentrations of amino acids should, theoretically, identify the limiting amino acid (McLaughlan, 1964), the alterations (Holt *et al.*, 1963), as well as the clinical manifestations, are likely to be indistinguishable from those seen in protein deficiency. Under experimental conditions, lysine deficiency has brought about decreased weight gain and decreased nitrogen retention without other manifestations (Graham *et al.*, 1969b), but has also resulted in hypoalbuminemia and hepatic steatosis when the diet is prolonged (Graham *et al.*, 1971).

Tryptophan deficiency merits special consideration. If it is moderate, if caloric intake is not excessive, and if adequate dietary niacin is available, manifestations of protein deficiency will appear. If deficiency is more severe and is accompanied by high energy intake, low niacin intake (Goldsmith, 1965), and possibly high leucine intake (Raghuramulu *et al.*, 1965), manifestations of pellagra will result. Truswell *et al.* (1968) have described a specific decrease in the plasma level of tryptophan.

In their studies of the essential amino acid requirements of infants, Snyderman and collaborators encountered specific manifestation of deficiency only when the diet lacked histidine or isoleucine; in all other cases, during the relatively short period of denial, no specific clinical signs appeared, although the plasma amino acids were altered in a fashion characteristic for each amino acid (Snyderman *et al.*, 1968). When infants less than 3 months of age were deprived of histidine, an eczematoid dermatitis appeared within a few days, but cleared promptly with restoration of the missing amino acid (Snyderman, 1965). In two of six infants, isoleucine withdrawal resulted in very pronounced loss of nitrogen, redness of the buccal mucosa, fissures at the corners of the mouth, tremors and twitching of the arms and legs, upward deviation of the eyes, and the lethargy and anorexia characteristic of other amino acid deficiency states (Snyderman *et al.*, 1964a). The marked central nervous system irritability disappeared promptly upon reincorporation of isoleucine in the diet. All of these manifestations, as well as the characteristic alteration of the plasma aminogram, resulted from complete withdrawal of each amino acid.

Effective treatment of phenylketonuria, an inborn error of phenylalanine metabolism, depends on the early reduction of phenylalanine intake. Special dietary products, practically devoid of phenylalanine, are used to lower plasma levels; and then milk protein is added cautiously to maintain a plasma phenylalanine concentration compatible with normal growth—low enough to prevent toxicity, yet high enough to prevent deficiency. When very low intakes of phenylalanine have been continued too long, persistent failure to gain weight, negative nitrogen balance, and, in some cases, death ensue. The skin is particularly sensitive: Alopecia occurs in mild deficiency, but in more severe cases a fiery red rash or exfoliative dermatitis develops, starting in the diaper area and spreading to the intertriginous folds. Lethargy, vomiting, and general misery are conspicuous (Woolf, 1963); and bone changes have been reported (Murdoch and Holman, 1964). Vitamin deficiencies, however, may have caused some of these symptoms, as well as other manifestations (Wilson and Clayton, 1962).

In the management of "maple syrup urine disease," an inborn error of the metabolism of the branched-chain amino acids, regulation of intakes of leucine, isoleucine, and valine, has prevented the adverse manifestations of the disease as well as any signs of deficiency of these amino acids (Snyderman et al., 1964b). When methionine intake was drastically curtailed in an attempt to reduce what was mistakenly thought to be a marked elevation in its plasma level, prompt and prolonged cessation of growth, but no other adverse manifestation, occurred.

Secondary and induced malnutrition   Management of pediatric disorders, notably diarrhea, often requires discontinuation of oral feedings and maintainenance of water, electrolyte, and glucose intake by the intravenous route. Reintroduction of oral feeding within 2 or 3 days, however, is usually advisable. When, because of necessity or excessive caution, feedings are withheld for longer periods, manifestations of acute protein deficiency may appear. Attempts at prolonged intravenous feeding, providing water, electrolytes, and glucose in generous amounts, can lead to hepatic steatosis, hypoalbuminemia, and salt retention, as well as development of edema. If amino acids are given intravenously, such manifestations can be prevented, assuming that there is no interference with protein synthesis (Wilmore and Dudrick, 1968).

Among the most "labile" of proteins are those of the pancreas and small intestine. Dietary protein deficiency results in a marked fall in the output of pancreatic enzymes and a failure of intestinal mucosal cells to regenerate (Bowie et al., 1965). The latter leads to a loss of absorptive surface and intestinal enzymes, particularly lactase. Full feeding may cause steatorrhea, decreased absorption, and diarrhea due to temporary disaccharide intolerance, particularly to lactose. Protein hydrolysates and nonlactose sugars will usually be well tolerated. Cautious introduction of diluted milk, allowing time for regeneration of mucosal cells, is also likely to meet with success.

A variety of intestinal malabsorption syndromes, resulting in inadequate digestion and absorption of protein, can cause protein deficiency. Incomplete digestion of starch and fat, with consequent caloric insufficiency, can lead to poor utilization of absorbed protein. A kwashiorkor-like picture has been described in infants with cystic fibrosis of the pancreas who were fed human milk or soybean milk (Fleisher et al., 1964). Changing to cow's milk, with its higher protein content, or adding pancreatic enzymes to the soybean milk, has resulted in satisfactory digestion of protein.

Prolonged use of adrenal cortical hormones profoundly affects protein metabolism. Although favoring the synthesis of the same vital proteins that are maintained during periods of caloric insufficiency (e.g., liver protein), these hormones cause increased catabolism and decreased synthesis of muscle and other tissue proteins, resulting in an unfavorable effect on growth and decreased nitrogen retention.

The role of infection   Infections play a significant role in the etiology of most cases of severe protein deficiency. Repeated episodes of gastrointestinal infection act in a variety of ways: Intake may decrease from anorexia and vomiting and the therapeutic withholding of food; the loss of nutrients may occur in diarrheal stools; utilization of protein may be impaired owing to systemic infection. Pyelonephritis, pulmonary tuberculosis, and other chronic bacterial infections interfere directly with protein synthesis and indirectly by increasing energy expenditure. Systemic viral infections profoundly alter the direction of protein synthesis in almost all cells toward the replication of virus protein. Measles is particularly important in the etiology of kwashiorkor (Scrimshaw, 1964).

The Adult

Most of the secondary or induced protein deficiency syndromes in children also occur in adults, but considerably less frequently. There are only four important syndromes in which evidences of protein depletion are prominent and in which dietary protein deficiency appears to play a role. These are hunger edema, kwashiorkor, pellagra, and nutritional liver disease. Only one of these, nutritional liver disease, is of any consequence today among adults in the United States.

Hunger edema   During and following World War I, in association with the extreme malnutrition existing in central Europe, observers came to recognize edema as a familiar accompaniment of semistarvation (Keys *et al.*, 1950; McCance, 1951). They soon demonstrated decreases in plasma protein concentrations; reductions in the albumin fraction of the serum were particularly prominent. During the 1920's and 1930's the concept that protein depletion and hunger or starvation edema were virtually synonymous was generally taught and accepted.

With the advent of World War II and the recurrence of mass semistarvation, the problem of the relation of dietary protein deficiency to edema again came into prominence; consequently, it received pointed study in its naturally occurring form and in human volunteers in the experimental laboratory (Keys *et al.*, 1950). Such investigations clearly indicated that the problem in man was not nearly so simple as had been thought. A number of important factors, some of which appeared to be more significant than dietary protein deficiency, were singled out and studied. In essence, it is now clear that in a given case of edema associated with energy and protein malnutrition, one or more of the following factors may operate: a general increase in body fluid that in part replaces lost fat and cells; salt intake; posture; tissue tensions; vascular permeability and introvascular pressure; renal function; hormonal effects; and intake of nonprotein nutrients, particularly the total energy intake. Varying degrees of protein depletion play a role, as well, in the deranged metabolism of water.

Kwashiorkor   The clinical picture of kwashiorkor, so often seen in small children, is seldom seen in adults, probably because of simultaneous severe energy deficiency and the lesser protein requirements of the mature individual. It has been described in association with chronic debilitating diseases, particularly chronic malabsorption, in adolescents with chronic infection, and in lactating women (Viteri *et al.*, 1964).

Pellagra   A relation between high-maize diets and pellagra has been recognized for more than a century. The low level of tryptophan in maize and the possibility that some of its niacin may not be available are of undoubted importance. Some rice diets, however, even lower in their tryptophan and niacin contents, are much less often pellagragenic. This suggests at least one other significant factor in maize, which may well be the unavailability of "bound" niacin (Goldsmith, 1964) or its elevated content of leucine (Raghuramulu *et al.*, 1965). Of primary importance in any pellagragenic diet, however, is the relation of total energy to niacin equivalents, either as the vitamin or as tryptophan (Goldsmith, 1965). In the United States, the alcoholic, who consumes little or no animal protein but relies on cornmeal or grits for his major supply of protein, will often develop pellagra. Those patients with carcinoid tumors of the intestine, who develop pellagra as a result of the markedly increased production of serotonin from tryptophan, are of special interest.

Nutritional liver disease   Increasing evidence has been accumulating on the role of nutritional factors in the pathogenesis of liver disease in man. It is difficult, however, to translate from carefully controlled laboratory observations to the complex biochemical and structural alterations that are observed in the human disease. The main benefit of the experimental studies has been to focus attention on nutritional liver disease; as a result, intensive biochemical and anatomical (biopsy) studies have been carried out during life and careful observations of the structure of the liver in health and disease have been reported. Undoubtedly, protein deficiency in man can lead to liver dysfunction, although the precise pathogenesis is not always clear.

Nutritional liver disease is often seen in alcoholics whose diets are poor; less frequently it occurs in individuals who are losing or destroying protein at an abnormal rate or are failing to synthesize adequate amounts of protein for a variety of reasons. The typical alcoholic may maintain an adequate energy intake, while his ingestion of essential nutrients, particularly protein, is exceedingly low. A considerable body of evidence suggests that there is an interplay between nutritional deficiency and a specific toxic effect of alcohol (Hartroft and Porta, 1966). The following syndromes, which merge together, may be encountered: simple fatty liver disease, fatty liver disease with hepatic failure, and nodular or Laennec's cirrhosis (Popper and Schaffner, 1957).

Simple fatty liver disease is frequently discovered incidentally in a known alcoholic or in an individual who, upon questioning, proves to be one. Except for an enlarged liver, which upon biopsy is shown to contain much fat, relatively little else may be found. If the individual is returned to a normal diet, the fat disappears from the liver and does not return, unless the diet again becomes inadequate.

If such a fatty liver persists for any length of time, one may expect to find evidence of hepatic functional changes—increased bromsulfophthalein retention, for example—and a rise in such indices as serum bilirubin and hypoproteinemia. At this stage, the individual may well begin to show acute liver failure, which appears to be aggravated by an infection. When severe hepatic insufficiency has developed, the course of the disease is difficult to reverse, and death frequently ensues. Autopsy will reveal an excessively fatty liver, which may or may not show fresh necroses and which usually displays relatively little excess connective tissue.

The third, cirrhotic, type develops initially as a fatty liver in which connective tissue slowly proliferates and interferes with circulation. This process further damages the liver cells. Moreover, since blood flow is obstructed, the increased pressure in the portal system leads to derangements elsewhere, particularly the formation of collateral venous channels and ascites. Liver failure may be the cause of death, or it may result from rupture of an esophageal varix, infection, or general debility.

In patients whose liver disease is of the first type (simple fatty liver disease), diet therapy is usually effective. A return to normal diet is sometimes effective before failure has occurred in the second group, though frequently all therapy is of no avail. When cirrhosis has developed, one may expect merely to arrest the development of the disease, since little can be done to alter the pattern of connective tissue in the liver. Hence adequate diet is indicated.

In nutritional liver disease, the role of protein is particularly difficult to evaluate. The decreased concentration of some plasma proteins, particularly albumin and the β-lipoproteins that are responsible for lipid transport out of the liver are indicative of protein disturbance. Some of this is doubtless due to insufficiency of amino acids for protein manufacture; much doubtless results from decreased formation of protein by a diseased liver.

Specific amino acid deficiencies   With the exception of the tryptophan deficiency that contributes to the development of pellagra and the possible role of methionine deficiency in nutritional liver disease, no other known instances of naturally occurring specific amino acid deficiencies can be identified, as such, in adult man. Manifestations of extreme irritability and fatigue, as well as a markedly negative nitrogen balance, developed promptly in all subjects who had isoleucine withdrawn from their diet under experimental conditions (Rose *et al.*, 1951; Swendseid and Dunn, 1956).

DEFICIENCY IN EXPERIMENTAL ANIMALS

Much knowledge concerning the role of protein and amino acids in nutrition has been obtained from studies of deficient states associated with deprivation of a particular material under study; information is available concerning experimental animals, particularly the rat, in which deficiency may be produced at will by dietary restriction. Caution must be exercised in drawing comparisons between artificially produced, single-nutrient deficiency states and naturally occurring, multiple-nutrient deficiency syndromes; only recently have close approximations of the latter been produced in experimental animals (Platt *et al.*, 1964; Pond *et al.*, 1965). Protein deficiency must be studied from two standpoints: quantity and quality. As for quantity, protein deficiency as a whole will be discussed; to consider quality, the effects of deprivation of single amino acids must be delineated.

When the protein intake of an experimental animal is restricted, certain biochemical and anatomical changes may be observed that reflect alterations in the protein composition of the cells and their extracellular products. Changes in the latter indicate biochemical lesions in the former, and thus serve as an index with which to assess cellular function.

If the experimental subject is growing, growth may be retarded or cease completely, or weight loss may occur. If the subject has attained its full growth, weight loss only may be encountered. If the animal has reached sexual maturity, reproductive activity, in both the male and the female, may be reduced or completely eliminated.

It is important to distinguish the precise effects of protein deficiency per se from those of energy restriction. Many recent studies have been carried out with this in mind and are particularly valuable for the light they throw on human protein depletion syndromes. The classical works in this field are *The Effects of Inanition and Malnutrition upon Growth and Structure*, by C. M. Jackson (1925), and *The Biology of Human Starvation*, by Keys *et al.* (1950).

A critical issue in experimentation is the matter of controls. Three methods have been utilized in the studies of dietary deficiency: (a) *ad libitum* feeding, (b) paired feeding, and (c) paired weight-gain feeding. Each method has its value, depending on the type of information sought. To compare structural alterations in the tissues, when it is important to detect the nonspecific effects of inanition, the paired weight-gain technique is preferable. Metabolic changes, the result of food restriction itself, must also be taken into account (Suttie, 1969).

Finally, it is important to recognize that the degree of deficiency that is produced may have an important bearing on the appearance and course of pathological changes. Acute, complete protein deficiency leads to relatively little change other than growth failure and atrophy of the tissues. On the other hand, less severe forms of deficiency, which may permit some biochemical activities as well as growth and reproduction of certain tissues, may provide much more information with respect to the metabolism and structure of cells and their extracellular products. In most animals a prompt decrease in the intake of a protein-free diet or one that is imbalanced with respect to its amino acid content will occur, thus imposing the effects of caloric deprivation on the changes observed. When animals are force-fed, the effects of deficiency of protein or specific amino acids can be more effectively observed. Although the rat will control its energy intake and thus make it difficult to produce a protein or amino acid deficiency, the pig, like the human, will continue to consume an inadequate diet.

Body Size and Composition

Animals fed on low-protein diets grow more slowly than do littermates on high-protein ones. The deprived animals consume less energy than a normally growing animal of the same age. If protein without extra calories is added to a low-protein diet, it promotes linear growth relative to body weight; the addition of extra carbohydrate without extra protein results in more gain in body weight than in bone growth; the latter may even be retarded. The organ weights of protein-deficient animals, as percentages of body weight, are more like those of younger animals of the same weight than those of a normal animal of the same age. In most forms of malnutrition, the proportion of water in the body is slightly increased. Body water content as high as 80 percent has been found in pigs fed a low-protein diet with supplemental carbohydrate (Platt *et al.*, 1964). Chickens fed diets of varying protein value demonstrated parallel variations in body composition, which, interestingly enough, were clearly reflected in the concentration of serum proteins, particularly the albumin fraction (Thomas and Combs, 1967). These observations should help establish the value of serum protein measurements in man as an index of protein nutriture.

Skeleton

As indicated above, growth of the protein-depleted organism may decrease sharply; hence, the skeletal tissues may reflect profound alterations (Follis, 1949; Frandsen *et al.*, 1954; Platt and Stewart, 1962). Depending on the severity of the deficient state, retardation of cell proliferation in the epiphyseal or costal cartilages will be retarded. These areas are a critical index of the nutritive status of the organism. The changes that take place in cartilage are not specific and differ in no way from those that occur as the result of restriction of energy intake or of deficiencies of any one of many nutrients. Because osteoblastic activity is greatly retarded, relatively little formation of periosteal and endosteal bone may be observed. In addition to the cessation of bone growth, the skeleton becomes rarefied. The latter sign also is observed in the adult whose endochrondral bone growth has ceased (Estramera and Armstrong, 1948). Evidences of disturbance in activity of the cells of the peridontal tissues and of the teeth themselves may be observed in protein-depleted animals (Hunter, 1950). Diets of cereal alone are more cariogenic than the same diets supplemented with high-quality animal protein (Dodds, 1964).

Connective Tissue Cells

The fibroblast, a cell closely related to those of cartilage and bone, is responsible for the elaboration of an important fibrous protein, collagen. One way of studying the activity of fibroblasts is to observe the formation of new collagen during the healing of artificially produced wounds. In severely protein-deficient animals, wound healing is markedly affected by the failure of collagen synthesis (Udupa *et al.*, 1956).

Reproductive Tract

In view of the effects on reproductive activity, the discovery of atrophy of tubules in the testes and impaired maturation of follicles in the ovaries of protein-deficient animals is not surprising (Nelson and Evans, 1953). Also, limited development of the various accessory sex organs would suggest that sex hormone production is reduced.

Endocrines

In association with changes in the gonads, alterations have been noted in other endocrine glands. The hypophysis and thyroid (Keys *et al.*, 1950), for example, may atrophy; on the other hand, the cortex of the adrenal may thicken (DaCosta and Clayton, 1952). Such changes are associated with an overall diminution in the amount of lymphoid tissue, including reduction in size of the thymus, lymph nodes, and spleen (Follis, 1958). Abnormalities in the anterior part of the hypophysis of protein–calorie deficient pigs (Platt *et al.*, 1964) include vacuolation of cells and loss of cytoplasm, as well as a crowding together of nuclei that makes identification of different cells difficult. The islet cells of the pancreas in protein–calorie-deficient pigs are crowded together, and the granules of the β-cells are decreased. The follicular cells of the thyroid glands are predominantly fat and elongated and have very few droplets within the cytoplasm, the colloid is mainly acidophilic, and few Aron vacuoles are present (Keys *et al.*, 1950).

Skin

Pigs maintained on protein–calorie-deficient diets have an unkempt appearance; their hair is sparse, fine, and long; their skin, dry and scaly, wrinkled and cracked. The dermis of deficient animals is disorganized, and the collagen is foamy and broken up with many, probably edematous, spaces. Fibroblasts are shrunken and histiocytes are increased. Similar changes have been reported in the rat, as well as changes in the amino acid composition of hair (Platt *et al.*, 1964).

Eye

McLaren (1958) fed rats from weaning to 6 months of age on diets with different protein values and found that, even on a cassava diet, the weight of the eyeball was only about 25 percent less than that of well-fed littermates. The proportions of water to dry matter did not vary. He also noted cataracts in pigs that had been maintained on diets low in protein for longer than 1 year (McLaren, 1959). The fibers of the lens were swollen and separated by vacuoles; later they degenerated, and the interfibrillar clefts filled with an amorphous deposit. Vascularization of the cornea appeared in rats given diets containing 3 percent or less of vegetable protein. The corneal epithelium reduced in thickness, and some keratinization of the superficial layers occurred.

Pancreas

The granular epithelial cells of the pancreas in protein-depleted animals exhibit a decrease in the amount of cytoplasm and a reduction in the number of granules that represent their secretory product (Grossman *et al.*, 1943). Monkeys tube-fed a protein-free diet exhibited atrophy of acinar cells in the pancreas (Deo *et al.*, 1965). Rabbits given a protein-deficient diet experienced acinar cell atrophy and a marked decrease in zymogen granules, in basal vacuoles, and in the various types of auto-phagic bodies, as well as a measurable loss of acid phosphate (Lazarus and Volk, 1965).

Gastrointestinal Tract

Epithelial cells in scrapings from the buccal mucosa of protein-deficient pigs and dogs stain poorly and are distorted and fragmented when compared with cells from normal animals (Squires, 1963). Animals given a protein-free diet (Nasset, 1964) lost nitrogen from the stomach and pancreas more rapidly than fasted animals. Both groups lost nitrogen from the liver at about the same rate, but those fed a protein-free diet lost much less nitrogen from the small intestine, and recovery of nitrogen was most rapid when an adequate diet was reinstated. This paradoxical situation is probably due to utilization by the small intestine of amino acids released from the gastric and pancreatic secretions during the initial phase of protein depletion. The intestines of protein–calorie-deficient animals appear to be of small diameter, and the mu-cuous and muscle coats are thin. The tips of the intestinal villi are in some instances bare of epithelial cells. Protein-deficient rhesus monkeys (Deo and Ramalingaswami, 1964) showed a marked decrease in absorption of vitamin B_{12}, in association with atrophy of the gastrointestinal mucosa, that was correctable by the administration of gastric juice.

Kidney

The renal cells in protein-deficient rats become atrophic, with no other particularly striking change (Jackson, 1925; Keys *et al.*, 1950).

Nervous Tissues

Changes in the appearance of the nerve and neuroglial cells have been observed in the spinal cord, medulla, and pons of protein–calorie deficient animals. The effects are less severe and more irregular at the higher levels of the brain. Alterations include chromatolysis, "foaming" of the cytoplasm, and an increase in the oligodendroglial nuclei, which are often clustered around damaged cells (Platt *et al.*, 1964). Activation of astrocytes, with an accompanying increase in the number and calibre of the astrocytic fibers, and, in the more severe cases, severe cell changes, neuronal loss, and heavy fibrous gliosis also occur. The neurofibrillar arrangement within the nerve cells becomes disorganized. Neither degeneration of myelinated fibers nor excess of fat are apparent.

The level of energy intake and the proportion of protein in the diet were found to influence the incidence of spontaneous chromophobe adenomas of the anterior pituitary of the male rat. Three purified diets, differing only in the energy–protein ratio, were employed. Each was provided throughout postweaning life, either *ad libitum* or in restricted but isocaloric amounts. Under restricted conditions, tumor prevalence related directly to the level of protein in the diet (Ross *et al.*, 1970).

In a study reported by Barnes *et al.* (1970), baby pigs were malnourished for a period of 8 weeks by restricting protein or energy intake in order to study behavioral changes that occurred during and long after nutritional rehabilitation. The nutritional conditions that caused the greatest change in behavioral development resulted from feeding a diet very low in protein from the 3rd through the 11th week of life.

Dobbing (1970) has summarized pertinent recent research dealing with the effects of protein and energy restriction on the central nervous system of growing experimental animals. Major emphasis centers on the effects of myelin formation and its lipid composition and on the role of the timing of malnutrition on cell number and size in the different regions of the brain. Of particular interest is the study of rehabilitated animals, showing the disappearance of compositional changes but the persistence of deficits in cell numbers and brain size, the latter being proportional to deficits in body size. Persistent functional and behavioral abnormalities have also been documented.

Muscle

Atrophy of cardiac, skeletal, and smooth muscle fibers appears in protein-depleted animals (Keys *et al.*, 1950). In protein-depleted rats, incorporation of amino acids increased in the liver but decreased in muscle. Concomitantly, the activity of amino acid activating enzymes increased in liver and decreased in muscle (Gaetani *et al.*, 1964). Also, increased sodium and decreased potassium content, similar to that of malnourished infants, has been found.

Another study compared the amino acid-incorporating activity of cell-free preparations from skeletal muscles of rats that were fed a protein-free diet for several days to that of rats that had received a meal adequate in protein following protein starvation. Both microsomal and ribosomal preparations from animals fed a protein-free diet for 4–6 days showed decreased amino acid-incorporating activity. Administration of a single meal of protein enhanced this activity, which was more pronounced with ribosomal (35–65 percent) than with microsomal preparations (15–40 percent). Incubation of the cell-free preparations with cell sap devoid of low molecular-weight components increased the ability of the microsomes and ribosomes to incorporate amino acids into protein. The monosomes were apparently reutilized for polypeptide formation. The data indicated that, although the activity of the animals' systems decrease after protein starvation, the functional sites for polypeptide formation remain fully intact (von der Decken and Omstedt, 1970).

Bone Marrow and Blood Cells

Among the most intensively studied effects of protein depletion is decrease in hemoglobin production and, hence, in the number of red blood cells (Madden *et al.*, 1945). The decrease in erythropoietic activity in protein-starved animals has been related to a decreased production of erythropoietin (Reissmann, 1964). Recent studies of special interest demonstrate that the anemia of malnutrition is often an adaptive process. hemoglobin mass being a function of metabolically active protoplasm (Stekel and Smith, 1969).

Liver

Microscopic examination of the cells of the liver from protein-deficient animals may reveal, in the first few days, more fat than normal (Kosterlitz, 1947). The RNA content falls subsequently; hence, the amount of basophilic material in the cytoplasm decreases. As might be expected, changes in the concentrations of numerous enzymes are demonstrable (Knox et al., 1956). Finally, decreases in the concentrations of fibrinogen, albumin, and some of the globulin fractions occur and may become marked (Miller, 1948). In the first month of protein depletion in the dog, the fractional loss of plasma proteins approximately equals that of whole body protein; but, in the second month, the loss of plasma proteins is proportionately much greater. Thus, plasma protein concentrations usually remain constant the first month and then drop sharply (Garrow, 1959).

The distribution of protein among the tissues of animals depleted of protein is the same as in the normal animal, except for a higher relative content in the brain of the depleted animal (Waterlow, 1959). When labeled methionine was administered to protein-depleted and normal animals, more of the label was found in the liver and less in the brain, muscle, carcass, and skin of depleted than of normal animals. In both groups, approximately equal amounts were found in visceral organs other than the liver. These studies suggest an increased rate of protein synthesis in the liver of the depleted animals. In animals fed a protein-free diet for prolonged periods of time, neutral glycerides and cholesterol in the liver were elevated for up to 8 weeks, then decreased to normal or below normal by 14 weeks (Williams and Hurlebaus, 1965).

Utilizing rats, investigators examined the hypothesis that fatty liver in protein malnutrition stems from a deficit in the availability of the protein moiety of low-density lipoproteins. Results of these studies strongly suggest this to be true, the impaired rate of synthesis of the lipoprotein causing triglycerides to accumulate in the liver (Flores et al., 1970). Also, a significant increase in the average DNA content of the liver cell nuclei and of the liver of rats on protein-deficient diets may be due to polyploidy, resulting from normal DNA synthesis but with blocked mitoses (Umana, 1965).

The livers of protein–energy-deficient pigs (Platt et al., 1964) are often yellow-brown, have well-marked lobules, and are extremely friable. Histological changes begin in the periportal area and spread, with increasing severity and duration of the protein deficiency, to the center of the lobules. The most obvious change is the vacuolated and angular appearance of the cells; the cytoplasm is broken up by the accumulation of glycogen and fat and loses much of its basophilia. The cells appear to be enlarged, and the sinusoids are relatively narrow. Some nuclei are enlarged and hypochromatic, others are small and hyperchromatic; most appear to be of normal size and to have only one, large, prominent nucleolus. Karyolysis occurs in the most severely affected livers, which also exhibit small necrotic areas. Fibrosis is not consistently present. Increasing the caloric intake of malnourished rats causes an increased deposition of fat; restricting the calorie intake, even when this involves overt protein deficiency, reduces the amount of fat and glycogen (Sidransky and Clark, 1961).

Immune Responses

A complicated, though important, consequence of protein deficiency is its effect on the immune reaction. Animals consuming protein-deficient diets have demonstrated decreases in antibody titer and alterations in the natural course of experimental infections (Cannon, 1945; Dubos and Schaedler, 1958). The response is even more complex when the infectious agent is a virus (Sprunt and Flanigan, 1956). In adult rats, titers of antibody but not of complement of properdin significantly decreased during protein depletion (Kenney et al., 1965).

Other workers have measured agglutinins and hemolysins in serum and antibody-forming cells in the spleens of adult male rats depleted of protein and then repleted with diets containing 5–10 percent of protein from rice or rice in combination with other proteins or crystalline amino acids. They then immunized the rats with sheep erythrocytes. Circulating antibodies generally paralleled counts of antibody-forming cells in the spleen. Increasing the concentration of dietary protein increased the weight of the spleen and the number of antibody-forming cells (Piedad-Pascual et al., 1970).

AMINO ACID DEFICIENCY IN EXPERIMENTAL ANIMALS

Studies of deficiencies of each of the essential amino acids in animals have shown few responses other than the nonspecific effects of protein deficiency already described. They have revealed alterations in all of the anatomical areas mentioned in the preceding sections.

Tryptophan

In addition to a host of nonspecific effects, tryptophan deprivation leads to the following: niacin deficiency, alopecia, and loss of pigment in incisor teeth (Albanese and Buschke, 1942) and necrosis of muscle and fatty liver (Adamstone and Spector, 1950). How many of these manifestations are due to nicotinic acid deficiency rather than to deprivation of tryptophan per se is not always clear. As for the fatty infiltration of the liver, tryptophan deficiency is one of several amino acid deficiencies leading to this change, which affects primarily the periportal parts of the hepatic lobule and contrasts with the centrolobular distribution of fat in choline deficiency. In the very young chicken, tryptophan deficiency inhibits growth of such tissues as connective tissue and cartilage (Gordon, 1964). Reference has already been made to the role of tryptophan in ribosomal aggregation (Fleck *et al.*, 1965).

Lysine

Deprivation of lysine leads to two specific changes in the rat: achromotrichia (Vohra and Kratzer, 1956) and periportal fatty liver (Singal *et al.*, 1953). Gray (1963) has demonstrated increased mortality from infections with anthrax bacillus in moderately deficient rats, possibly because of decreased function of the reticuloendothelial system. In young rats force-fed a lysine-deficient diet, the following occurred: Increased lipid and glycogen in the liver; atrophy of the pancreas, salivary glands, and spleen; and no change in the stomach (Sidransky and Verney, 1964).

Isoleucine

Studies of isoleucine-deficient rats revealed necrosis of the skeletal muscle (Scott, 1956). In young rats force-fed isoleucine-deficient diets (Sidransky and Verney, 1964) changes resembled those seen in histidine, phenylalanine, threonine, and valine deficiency: increase in fat and glycogen content in the liver; some pancreatic atrophy; and some shrinking of salivary glands, stomach, and spleen. Observations in other similar experiments (Lyman *et al.,* 1964a) included an increase in acetate incorporation, a decrease in linoleic acid content, and an increase in the palmitic acid content of the liver. Glycogen also markedly increased. These studies, in contrast to those on choline–methionine deficiency, in which the defect appeared to be a block in fat transport from the liver, suggest that fatty acid synthesis increased in the liver during isoleucine deficiency. Further studies have revealed increased concentration of liver triglyceride; increased proportion of palmitic, stearic, and oleic acids; decreased proportion of linoleic and arachidonic acids; no impairment of lipoprotein synthesis; and some impairment of removal of liver triglycerides (Lyman *et al.,* 1964b). Liver polysome patterns of rats fed isoleucine- or threonine-deficient diets resemble those induced by tryptophan-deficient diets (Pronczuk *et al.*, 1970).

Threonine

In young rats force-fed a threonine-deficient diet, the same changes were observed as in those fed an isoleucine-deficient diet (Sidransky and Verney, 1964). The livers of rats fed diets lacking threonine showed an accelerated rate of incorporation of amino acids into protein, which is related to enhanced ribosomal activity (Sidransky *et al.*, 1964).

Methionine

Methionine is a precursor of cystine; lack of it will give rise to methionine- and cystine-deficient states unless cystine is added to the diet. The presence of adequate dietary cystine aggravates the fatty liver caused by combined choline and methionine deficiency (Follis, 1958), probably due to the stimulation of growth. The increased liver triglycerides and decreased serum lipids in choline–methionine deficiency are probably due to decreased ability to transport lipid from liver to the blood. In the presence of choline, methionine deficiency results only in depressed formation of hemoglobin and plasma proteins. With suboptimal amounts of methionine in the diet and no added cystine, acute necrosis of the liver occurs (Glynn *et al.*, 1945). This situation is made worse by lack of tocopherol (György and Goldblatt, 1949) and of selenium (Schwarz and Foltz, 1958).

In young rats force-fed a methionine-devoid diet, liver lipids increase and a slight decrease in body weight and protein content of skeletal muscle occurs (Sidransky and Verney, 1964). Weight loss, unkempt fur, hyperirritability, and porphyrin staining about whiskers and forepaws are evident in adult male and female rats force-fed a methionine-deficient diet (Lyman *et al.*, 1964b). Observers also noted markedly increased liver triglycerides and markedly decreased glycogen, but found little evidence of effect on liver weight, phospholipids, nitrogen, and cholesterol. The total liver lipid of the female, however, was markedly increased; that of the male only questionably so. The female rats exhibited a questionable decrease in incorporation of acetate into the liver lipid and, as compared to the male, a greater increase in the proportion of linoleic acid (probably the result of accumulation of dietary fat or fatty acids from adipose tissue) and decrease in palmitic acid. Also, serum cholesterol and phospholipids of the female, but not of the male, were decreased. In contrast to isoleucine deficiency, methionine lack caused no change in acetate incorporation.

Valine

Convulsions have been reported in rats whose diets were deficient in valine (Rose and Epstein, 1939; Ferraro and Roizin, 1947). Valine-deficient diets force-fed to young rats produced changes similar to those seen in rats force-fed histidine-, isoleucine-, phenylalanine-, and threonine-deficient diets (Sidransky and Verney, 1964).

REFERENCES

Adamstone, F. B., and H. Spector. 1950. Tryptophan deficiency in the rat; histologic changes induced by forced feeding of an acid-hydrolyzed casein diet. Arch. Pathol. 49:173–184.

Albanese, A. A., and W. Buschke. 1942. On cataract and certain other manifestations of tryptophane deficiency in rats. Science 95:584–586.

Alleyne, G. A. O., D. J. Millward, and G. H. Scullard. 1970. Total body potassium, muscle electrolytes, and glycogen in malnourished children. J. Pediatr. 76:75–81.

Ashley, J. H., and H. Fisher. 1967. Protein reserves and muscle constituents of protein-depleted and repleted cocks. Br. J. Nutr. 21:661–670.

Ashworth, A. 1969. Metabolic rates during recovery from protein–calorie malnutrition: The need for a new concept of specific dynamic action. Nature 223:407–409.

Barbezat, G. O., and J. D. L. Hansen. 1968. The exocrine pancreas and protein-calorie malnutrition. Pediatrics 42:77–92.

Barnes, R. H., A. U. Moore, and W. G. Pond. 1970. Behavioral abnormalities in young adult pigs caused by malnutrition in early life. J. Nutr. 100:149–155.

Bowie, M. D., G. L. Brinkman, and J. D. L. Hansen. 1965. Acquired disaccharide intolerance in malnutrition. J. Pediatr. 66:1083–1091.

Bradfield, R. B., M. A. Bailey, and A. Cordano. 1968. Hair-root changes in Andean Indian children during marasmic kwashiorkor. Lancet 2:1169–1170.

Bradfield, R. B., A. Cordano, and G. G. Graham. 1969. Hair-root adaptation to marasmus in Andean Indian children. Lancet 2:1395–1397.

Brunser, O., A. Reid, F. Monckeberg, A. Maccioni, and I. Contreras. 1966. Jejunal biopsies in infant malnutrition, with special reference to mitotic index. Pediatrics 38:605–612.

Cannon, P. R. 1945. The relation of protein metabolism to antibody production and resistance to infection. Adv. Protein Chem. 2:135–154.

Chase, H. P., and H. P. Martin. 1970. Undernutrition and child development. N. Engl. J. Med. 282:933–939.

Cheek, D. B., ed. 1968. Human growth; body composition, cell growth, energy and intelligence. Lea and Febiger, Philadelphia. 781 pp.

Cheek, D. B., D. E. Hill, A. Cordano, and G. G. Graham. 1970. Malnutrition in infancy: Changes in muscle and adipose tissue before and after rehabilitation. Pediatr. Res. 4:135–144.

DaCosta, E., and R. Clayton. 1952. Histological and biochemical changes in the rat adrenal after dietary restriction and rehabilitation. Am. J. Physiol. 171:717. (A)

Deo, M. G., and V. Ramalingaswami. 1964. Adsorption of Co-58 labeled cyanocobalamin in protein deficiency. An experimental study in the Rhesus monkey. Gastroenterology 46:167–174.

Deo, M. G., S. K. Sood, and V. Ramalingaswami. 1965. Experimental protein deficiency: Pathological features in the Rhesus monkey. Arch. Pathol. 80:14–23.

Dobbing, J. 1970. Undernutrition and the developing brain. The relevance of animal models to the human element. Am. J. Dis. Child. 120:411–415.

Dodds, M. L. 1964. Protein and lysine as factors in the cariogenicity of a cereal diet. J. Nutr. 82:217–223.

Dubos, R. J., and R. W. Schaedler. 1958. Effect of dietary proteins and amino acids on susceptibility to bacterial infections. J. Exp. Med. 108:69–81.

Dugdale, A. E., S. T. Chen, and G. Hewitt. 1970. Patterns of growth and nutrition in children. Am. J. Clin. Nutr. 23:1280–1287.

Estramera, H. R., and W. D. Armstrong. 1948. Effect of protein intake on the bones of mature rats. J. Nutr. 35:611–618.

Ferraro, A., and L. Roizin. 1947. Essential amino-acid deficiency, clinico-pathologic findings in rats. II. Valine. J. Neuropathol. Exp. Neurol. 6:383–390.

Fleck, A., J. Shepherd, and H. N. Munro. 1965. Protein synthesis in rat liver: Influence of amino acids in diet on microsomes and polysomes. Science 150:628–629.

Fleisher, D. S., A. M. DiGeorge, L. A. Barness, and D. Cornfeld. 1964. Hypoproteinemia and edema in infants with cystic fibrosis of the pancreas. J. Pediatr. 64:341–348.

Flores, H., W. Sierralta, and F. Monckeberg. 1970. Triglyceride transport in protein-depleted rats. J. Nutr. 100:375–379.

FNB (Food and Nutrition Board, National Research Council). 1970. Maternal nutrition and the course of pregnancy. National Academy of Sciences, Washington, D.C. 241 pp.

Follis, R. H., Jr. 1949. Nutrition and bone disease. J. Mt. Sinai Hosp. N.Y. 16:1–13.

Follis, R. H., Jr. 1958. Deficiency disease. C. C. Thomas, Springfield, Ill. 577 pp.

Frandsen, A. M., M. M. Nelson, E. Sulon, H. Becks, and H. M. Evans. 1954. The effects of various levels of dietary protein on skeletal growth and endochondral ossification in young rats. Anat. Rec. 119:247–265.

Frenk, S., J. Metcoff, F. Gomez, R. Ramos-Galvan, J. Cravioto, and I. Antonowicz. 1957. Intracellular composition and homeostatic mechanisms in severe chronic infantile malnutrition. II. Composition of tissues. Pediatrics 20:105–120.

Gaetani, S., A. M. Paolucci, M. A. Spadoni, G. Tomassi. 1964. Activity of amino acid-activating enzymes in tissues from protein-depleted rats. J. Nutr. 84:173–178.

Garn, S. M., C. G. Rohmann, M. Béhar, F. Viteri, and M. A. Guzman. 1964. Compact bone deficiency in protein–calorie malnutrition. Science 145:1444–1445.

Garrow, J. S. 1959. The effect of protein depletion on the distribution of protein synthesis in the dog. J. Clin. Invest. 38:1241–1250.

Garrow, J. S., and M. C. Pike. 1967. The long-term prognosis of severe infantile malnutrition. Lancet 1:1–4.

Glynn, L. E., H. P. Himsworth, and A. Neuberger. 1945. Pathological states due to deficiency of the sulfur-containing amino acids. Br. J. Exp. Pathol. 26:326–337.

Goldsmith, G. A. 1964. The B vitamins: Thiamine, riboflavin, niacin. Pages 161–206 *in* G. H. Beaton and E. W. McHenry, eds. Nutrition, a comprehensive treatise. Vol. II. Academic Press, New York.

Goldsmith, G. A. 1965. Niacin: Antipellagra factor, hypocholesterolemic agent. Model of nutrition research yesterday and today. J. Am. Med. Assoc. 194:167–173.

Gordon, R. S. 1964. Growth stasis and tryptophan deficiency in the very young chicken. Nature 203:320–321.

Graham, G. G. 1967. Effect of infantile malnutrition on growth. Fed. Proc. 26:139–143.

Graham, G. G. 1968. The later growth of malnourished infants; effects of age, severity, and subsequent diet. Pages 301–316 *in* R. A. McCance and E. M. Widdowson, eds. Calorie deficiencies and protein deficiencies. J. & A. Churchill, London.

Graham, G. G., and B. Adrianzen. 1970. Growth, inheritance and environment. Pediatr. Res. 4:375. (A)

Graham, G. G., and A. Cordano. 1969. Copper depletion and deficiency in the malnourished infant. John Hopkins Med. J. 124:139–150.

Graham, G. G., A. Cordano, R. M. Blizzard, and D. B. Cheek. 1969a. Infantile malnutrition: Changes in body composition during rehabilitation. Pediatr. Res. 3:579–589.

Graham, G. G., R. P. Placko, G. Acevedo, E. Morales, and A. Cordano. 1969b. Lysine enrichment of wheat flour: Evaluation in infants. Am. J. Clin. Nutr. 22:1459–1468.

Graham, G. G., E. Morales, A. Cordano, and R. P. Placko. 1971. Lysine enrichment of wheat flour: Prolonged feeding of infants. Am. J. Clin. Nutr. 24:200–206.

Gray, I. 1963. Lysine deficiency and host resistance to anthrax. J. Exp. Med. 117:497–508.

Grossman, M. I., H. Greenard, and A. C. Ivy. 1943. Effect of dietary composition on pancreatic enzymes. Am. J. Physiol. 138:676–682.

György, P., and H. Goldblatt. 1949. Further observations on the production and prevention of dietary hepatic injury in rats. J. Exp. Med. 89:245–268.

Hartroft, W. S., and E. A. Porta. 1966. Experimental alcoholic hepatic injury. Nutr. Rev. 24:97–101.

Holt, L. E., Jr., S. E. Snyderman, P. M. Norton, E. Roitman, and J. Finch. 1963. The plasma aminogram in kwashiorkor. Lancet 2:1343–1348.

Hunter, H. A. 1950. Hypoproteinemia in relation to the dental tissues. J. Dent. Res. 29:73–86.

IUNS (International Union of Nutritional Sciences, Committee on Procedures for Appraisal of Protein–Calorie Malnutrition). 1970. Assessment of protein nutritional status. Am. J. Clin. Nutr. 23:807–819.

Jackson, C. M. 1925. The effects of inanition and malnutrition upon growth and structure. Blakiston, Philadelphia. 616 pp.

James, W. P. T., and H. G. Coore. 1970. Persistent impairment of insulin secretion and glucose tolerance after malnutrition. Am. J. Clin. Nutr. 23:386–389.

James, W. P. T., and A. M. Hay. 1968. Albumin metabolism: Effects of the nutritional state and the dietary protein intake. J. Clin. Invest. 47:1958–1972.

Jelliffe, D. B., G. Bras, and K. L. Stuart. 1954. Kwashiorkor and marasmus in Jamaican infants. West Indian Med. J. 3:43–55.

Kenney, M. A., L. Arnich, E. Mar, and C. E. Roderuck. 1965. Influence of dietary protein on complement, properdin, and hemolysin in adult protein-depleted rats. J. Nutr. 85:213–220.

Keys, A., J. Brozek, A. Henschel, O. Mickelson, and H. L. Taylor. 1950. The biology of human starvation. 2 Vols. University of Minnesota Press, Minneapolis.

Kimura, K. 1967. A consideration of the secular trend in Japanese for height and weight by a graphic method. Am. J. Phys. Anthropol. 27:89–94.

Knox, W. E., V. H. Auerbach, and E. C. C. Lin. 1956. Enzymatic and metabolic adaptations in animals. Physiol. Rev. 36:164–254.

Kosterlitz, H. W. 1947. The effects of changes in dietary protein in the composition and structure of the liver cell. J. Physiol. 106:194–210.

Lazarus, S. S., and B. W. Volk. 1965. Ultrastructure and acid phosphate distribution in the pancreas of rabbits. A comparison of alterations following protein deficient and semistarvation diets. Arch. Pathol. 80:135–147.

Lyman, R. L., C. R. Cook, and M. A. Williams. 1964a. Liver lipid accumulation in isoleucine-deficient rats. J. Nutr. 82:432–438.

Lyman, R. L., S. Thenen, and C. R. Cook. 1964b. Methionine deficiency and fatty liver in male and female rats. Proc. Soc. Exp. Biol. Med. 117:696–699.

McCance, R. A. 1951. The history, significance and aetiology of hunger oedema. in Studies of undernutrition, Wuppertal, 1946–49. Med. Res. Counc. (G.B.) Spec. Rep. Ser. No. 275:21–82.

McLaren, D. S. 1958. Growth and water content of the eyeball of the albino rat in protein deficiency. Br. J. Nutr. 12:254–259.

McLaren, D. S. 1959. The eye and related glands of the rat and pig in protein deficiency. Br. J. Ophthalmol. 43:78–87.

McLaughlan, J. M. 1964. Blood amino acid studies. V. Determination of the limiting amino acid in diets. Can. J. Biochem. 42:1353–1360.

Madden, S. C., F. W. Anderson, J. C. Donovan, and G. H. Whipple. 1945. Plasma protein production influenced by amino acid mixtures and lack of essential amino acids. J. Exp. Med. 82:77–92.

Miller, L. L. 1948. Changes in rat liver enzyme activity with acute inanition: Relation of loss of enzyme activity to liver protein loss. J. Biol. Chem. 172:113–121.

Montgomery, R. D. 1961. Magnesium balance studies in marasmic kwashiorkor. J. Pediatr. 59:119–123.

Munro, H. N. 1964. General aspects of the regulation of protein metabolism by diet and by hormones. Pages 381–481 in H. N. Munro and J. B. Allison, eds. Mammalian protein metabolism. Vol. I. Academic Press, New York.

Murdoch, M. M., and G. H. Holman. 1964. Roentgenologic bone changes in phenylketonuria. Relation to dietary phenylalanine and serum alkaline phosphatase. Am. J. Dis. Child. 107:523–532.

Nasset, E. S. 1964. The role of the digestive tract in protein metabolism. Am. J. Dig. Dis. 9:175–190.

Nelson, N. M., and H. M. Evans. 1953. Relation of dietary protein levels to reproduction in the rat. J. Nutr. 51:71–84.

Piedad-Pascual, F., L. Arnich, and M. A. Kenney. 1970. Rice-based protein mixtures: Immune and hepatic responses in rats. J. Nutr. 100:389–396.

Platt, B. S., and R. J. C. Stewart. 1962. Transverse trabeculae and osteoporosis in bones in experimental protein–calorie deficiency. Br. J. Nutr. 16:483–495.

Platt, B. S., C. R. R. Heard, and R. J. C. Stewart. 1964. Experimental protein, calorie deficiency. Pages 445–521 in H. N. Munro and J. B. Allison, eds. Mammalian protein metabolism. Vol. II. Academic Press, New York.

Pond, W. G., R. H. Barnes, R. B. Bradfeld, E. Kwong, and L. Krook. 1965. Effect of dietary energy intake on protein deficiency symptoms and body composition of baby pigs fed equalized but suboptimal amounts of protein. J. Nutr. 85:57–66.

Popper, H. P., and F. Schaffner. 1957. Liver: Structure and function. Blakiston, New York. 777 pp.

Prader, A., J. M. Tanner, and G. A. von Harnack. 1963. Catch-up growth following illness or starvation; an example of developmental canalization in man. J. Pediatr. 62:646–659.

Pronczuk, A. W., Q. R. Rogers, and H. N. Munro. 1970. Liver polysome patterns of rats fed amino acid imbalanced diets. J. Nutr. 100:1249–1258.

Raghuramulu, N., B. S. Narasinga Rao, and C. Gopalan. 1965. Amino acid imbalance and tryptophan–niacin metabolism. I. Effect of excess leucine on the urinary excretion of tryptophan–niacin metabolites in rats. J. Nutr. 86:100–106.

Reissmann, K. R. 1964. Protein metabolism and erythropoiesis. II. Erythropoietin formation and erythroid responsiveness in protein-deprived rats. Blood 23:146–153.

Rose, W. C., and S. H. Epstein. 1939. The dietary indispensability of valine. J. Biol. Chem. 127:677–684.

Rose, W. C., W. J. Haines, and D. T. Warner. 1951. The amino acid requirements of man. III. The role of isoleucine: Additional evidence concerning histidine. J. Biol. Chem. 193:605–612.

Ross, M. H. 1964. Nutrition, disease and length of life. Pages 90–107 in G.E.W. Wolstenholme and M. O'Connor, eds. Diet and Bodily Constitution. CIBA Foundation, Study Group No. 17. Little, Brown and Co., Boston.

Ross, M. H., G. Bras, and M. S. Ragbeer. 1970. Influence of protein and calorie intake upon spontaneous tumor incidence of the anterior pituitary gland of the rat. J. Nutr. 100:177–189.

Schwarz, K., and C. M. Foltz. 1958. Factor 3 activity of selenium compounds. J. Biol. Chem. 233:245–251.

Scott, E. B. 1956. Histopathology of amino acid deficiencies. V. Isoleucine. Proc. Soc. Exp. Biol. Med. 92:134–140.

Scrimshaw, N. S. 1964. Protein deficiency and infective disease. Pages 569–592 in H. N. Munro and J. B. Allison, eds. Mammalian protein metabolism. Vol. II. Academic Press, New York.

Sidransky, H., and S. Clark. 1961. Chemical pathology of acute amino acid deficiencies. IV. Influence of carbohydrate intake on the morphologic and biochemical changes in young rats fed threonine- or valine-deficient diets. Arch. Pathol. 72:468–479.

Sidransky, H., and E. Verney. 1964. Chemical pathology of acute amino acid deficiencies. VII. Morphologic and biochemical changes in young rats force-fed arginine-, leucine-, isoleucine-, or phenylalanine-devoid diets. Arch. Pathol. 78:134–148.

Sidransky, H., T. Staehelin, and E. Verney. 1964. Protein synthesis enhanced in the liver of rats force-fed a threonine-devoid diet. Science 146:766–768.

Singal, S. A., S. J. Hazan, V. P. Sydenstricker, and J. M. Littlejohn. 1953. The production of fatty livers in rats on threonine- and lysine-deficient diets. J. Biol. Chem. 200:867–874.

Snyderman, S. E. 1965. An eczematoid dermatitis in histidine deficiency. J. Pediatr. 66:212–215.

Snyderman, S. E., A. Boyer, P. M. Norton, E. Roitman, and L. E. Holt, Jr. 1964a. The essential amino acid requirements of infants. IX. Isoleucine. Am. J. Clin. Nutr. 15:313-321.

Snyderman, S. E., P. M. Norton, E. Roitman, and L. E. Holt, Jr. 1964b. Maple syrup urine disease, with particular reference to dietotherapy. Pediatrics 34:454-472.

Snyderman, S. E., L. E. Holt, Jr., P. M. Morton, and E. Roitman. 1968. Effect of high and low intakes of individual amino acids on the plasma aminograms. Pages 19-31 in J. H. Leathem, ed. Protein nutrition and free amino acid patterns. Rutgers University Press, New Brunswick, N.J.

Sprunt, D. H., and C. C. Flanigan. 1956. The effect of malnutrition on the susceptibility of the host to viral infection. J. Exp. Med. 104:687-706.

Squires, B. T. 1963. The buccal mucosa in protein-calorie deficiency in the pig. Br. J. Nutr. 17:303-307.

Standard, K. L., V. G. Wills, and J. C. Waterlow. 1959. Indirect indicators of muscle mass in malnourished children. Am. J. Clin. Nutr. 7:271-279.

Stekel, A., and N. J. Smith. 1969. Hematologic studies of severe undernutrition of infancy. II. Erythropoietic response to phlebotomy by calorie-deprived pigs. Pediatr. Res. 3:338-345.

Stoch, M. B., and P. M. Smythe. 1967. The effect of undernutrition during infancy on subsequent brain growth and intellectual development. S. Afr. Med. J. 41: 1027-1030.

Suttie, J. W. 1969. Effect of a dietary fluoride on the pattern of food intake in the rat and the development of a programmed pellet dispenser. J. Nutr. 96:529-535.

Swendseid, M. E., and M. S. Dunn. 1956. Amino acid requirements of young women based on nitrogen balance data. II. Studies on isoleucine and on minimum amounts of the eight essential amino acids fed simultaneously. J. Nutr. 58:507-517.

Thomas, O. P., and G. F. Combs. 1967. Relationship between serum protein level and body composition in the chick. J. Nutr. 91:468-472.

Truswell, A. S., J. D. L. Hansen, and P. Wannenburg. 1968. Plasma tryptophan and other amino acids in pellagra. Am. J. Clin. Nutr. 21:1314-1320.

Truswell, A. S., J. D. L. Hansen, C. E. Watson, and P. Wannenburg. 1969. Relation of serum lipids and lipoproteins to fatty liver in kwashiorkor. Am. J. Clin. Nutr. 22:568-576.

Udupa, K. N., J. F. Woessner, and J. E. Dunphy. 1956. The effect of methionine on the production of mucopolysaccharides and collagen in healing wounds of protein-depleted animals. Surg. Gynecol. Obstet. 102:639-645.

Umana, R. 1965. Effects of protein malnutrition on the DNA content of rat liver. J. Nutr. 85:169-173.

Viteri, F. E., and J. Alvarado. 1970. The creatinine height index: Its use in the estimation of the degree of protein depletion and repletion in protein calorie malnourished children. Pediatrics 46:696-707.

Viteri, F., M. Béhar, and G. Arroyave. 1964. Clinical aspects of protein malnutrition. Pages 523-568 in H. N. Munro and J. B. Allison, eds. Mammalian protein metabolism. Vol. II. Academic Press, New York.

Vohra, P., and F. H. Kratzer. 1956. Graying of hair in rats fed on ration deficient in lysine. Science 124:1145.

von der Decken, A., and P. T. Omstedt. 1970. Protein feeding of rats after starvation: Incorporation of amino acids into polypeptide by skeletal muscle ribosomes. J. Nutr. 100:623-630.

Waterlow, J. C. 1959. Effects of protein depletion on the distribution of protein synthesis. Nature 184:1875–1876.

Waterlow, J. C. 1968. The adaptation of protein metabolism to low protein intakes. Pages 61–73 *in* R. A. McCance and E. M. Widdowson, eds. Calorie deficiencies and protein deficiencies. J. & A. Churchill, London.

Waterlow, J. C., and V. G. Wills. 1960. Balance studies in malnourished Jamaican infants. I. Absorption and retention of nitrogen and phosphorus. Br. J. Nutr. 14:183–205.

Williams, J. N., Jr., and A. J. Hurlebaus. 1965. Response of the liver to prolonged protein depletion. V. Neutral glycerides and cholesterol; production of fatty livers by certain amino acids in a protein-free ration. J. Nutr. 85:73–81.

Wilmore, D. W., and S. J. Dudrick. 1968. Growth and development of an infant receiving all nutrients exclusively by vein. J. Am. Med. Assoc. 203:860–864.

Wilson, K. M., and B. E. Clayton. 1962. Importance of choline during growth, with particular reference to synthetic diets in phenylketonuria. Arch. Dis. Child. 37:565–577.

Winick, M., and P. Rosso. 1969. The effect of severe early malnutrition on cellular growth of human brain. Pediatr. Res. 3:181–184.

Woolf, L. I. 1963. Inherited metabolic disorders; errors of phenylalanine and tyrosine metabolism. Adv. Clin. Chem. 6:97–230.

A.E. HARPER

Effects of Disproportionate
Amounts of Amino Acids

The proportions of amino acids in diets usually differ from the proportions required by the body; yet adverse effects other than low efficiency of nitrogen utilization, are uncommon. Nevertheless, adverse effects have been observed in experimental animals consuming diets containing disproportionate amounts of amino acids, usually much greater than would be encountered in nature. Such effects might be anticipated when homeostatic mechanisms regulating amino acid concentrations in body fluids are deficient or defective, or are artificially overloaded. This situation could well occur in human subjects whose ability to handle an amino acid load is impaired because of liver damage, malnutrition, or a genetic defect of amino acid metabolism. Therefore, observations from studies on experimental animals and the limited information from human studies have been examined in an effort to assess the likelihood of similar effects resulting from the use of amino acid supplements.

Information about adverse effects from ingestion of diets containing disproportionate amounts of amino acids has been reviewed recently (Harper *et al.*, 1970), as has information about the pharmacology (Milne, 1968) and toxicology of amino acids (Harper, 1966, 1973). The 1970 review provides considerable detail about animal studies and an extensive bibliography; the others place more emphasis on observations of human subjects. The material presented here is condensed in large measure from two of those (Harper *et al.*, 1970; Harper, 1973).

Adverse effects, ranging from moderate depressions of food intake and growth to the development of pathological lesions and low survival rates, have occurred in animals consuming an excessive amount of a given amino acid, the most severe effects occurring in young growing animals fed a low-protein diet containing an inordinately large amount of one amino acid. Depressions in food intake and growth, usually less severe than those resulting from excessive intakes of particular amino acids, have also been observed in animals fed diets low in protein and containing a more general disproportion of amino acids in relation to the requirements for normal growth and maintenance. The severity of the adverse effects varies not only with the nature and degree of the amino acid disproportion but also with the nutritional adequacy of the diet and with the age and physiologic state of the animal.

In normal human subjects who have been administered large doses of individual amino acids, any adverse effects have usually been mild. However, mental deficiency is commonly associated with metabolic defects of amino acid metabolism that result in prolonged accumulation of an amino acid in body fluids.

TYPES OF ADVERSE EFFECTS

Adverse effects in animals resulting from the ingestion of diets containing disproportionate amounts of amino acids have been attributed to (a) toxicities, (b) antagonisms, and (c) imbalances (Elvehjem, 1956; Harper, 1956). Because different authors have used these terms in different ways, the following definitions will apply to this report.

Toxicity

The term "amino acid toxicity" includes adverse effects of varying degrees resulting from ingestion of large quantities of individual amino acids. For example, tyrosine, if ingested in large amounts by young growing rats, is clearly toxic, causing severe eye and paw lesions, and, if ingested in great excess, is lethal (Schweizer, 1947); whereas threonine even in great excess, causes only moderate retardation of growth and depression of food intake (Sauberlich, 1961).

Antagonisms

Depressions of growth caused by ingestion of excessive amounts of naturally occurring amino acids and alleviated by supplements of structurally similar amino acids are attributed to amino acid antagonisms.

Two examples have been described: one among leucine, isoleucine, and valine (Harper *et al.*, 1955), and another between lysine and arginine (Jones, 1964). The bases for these antagonisms have not been clearly established, therefore it should not be assumed that the term implies classical antimetabolite action.

Imbalances

Adverse effects from surpluses of essential amino acids other than the one that is limiting for growth or maintenance are commonly attributed to amino acid imbalances (Harper, 1967). This term has been used most specifically by investigators of nicotinic acid deficiency, who attributed to an amino acid imbalance the depressed growth rate of rats fed a low-protein, nicotinic acid-deficient diet to which a quantity of a tryptophan-deficient protein had been added (Elvehjem and Krehl, 1947, 1955). Others have reported many examples of growth depressions caused by additions of amino acid-deficient proteins or nutritionally incomplete mixtures of amino acids to low-protein diets that contain adequate quantities of all vitamins and minerals; these depressions are readily prevented by a small supplement of the growth-limiting amino acids (Harper, 1958a). In some studies of animals fed low-protein diets, the amounts of amino acids observed to cause adverse effects attributable to amino acid imbalances have represented as little as one-fifth of the requirement.

It is essential to distinguish clearly between imbalances and deficiencies; investigations of deficiencies deal with the effects of an inadequate intake of an amino acid, unlike imbalances, in which concern is with the effects of surpluses of amino acids. An imbalance is distinct from toxicity in that the latter term applies to conditions in which an adverse effect is due to a large surplus of an individual amino acid. To create imbalances, the total quantity of amino acids added may be greater than the quantities causing toxicity, but usually no single amino acid is included in the diet in an amount that, by itself, would be considered toxic.

EFFECTS OF EXCESSIVE INTAKES OF INDIVIDUAL AMINO ACIDS

Aromatic Amino Acids

Effects of excessive intakes of phenylalanine and tyrosine have received more attention than have effects of excessive intakes of other amino acids because of the great interest in developing experimental animal models of the genetic defects of aromatic amino acid metabolism, alcaptonuria, phenylketonuria, and tyrosyluria.

Inclusion of 4 percent or more of phenylalanine in an adequate diet (Hier *et al.,* 1944) or in a diet inadequate in protein (Wretlind, 1952) depresses growth and food intake of the rat, the severity of the growth depression increasing with increasing phenylalanine content of the diet (Benton *et al.,* 1956a; Kerr and Waisman, 1967). Growth retardation lessens as the quality or the quantity of protein in the diet increases (Benton *et al.,* 1956a; Sauberlich, 1961; Harper *et al.,* 1966).

Elevation of plasma phenylalanine and tyrosine concentrations in various species of animals (Sauberlich, 1961; Waisman and Harlow, 1965; Kerr and Waisman, 1967) and in children (Snyderman *et al.,* 1968) fed diets high in phenylalanine has been well documented, as has elevated plasma concentration and urinary excretion of phenylpyruvate (Auerbach *et al.,* 1958; Goldstein, 1961). Evidence that a high phenylalanine intake causes behavioral changes in rats and monkeys (Waisman and Harlow, 1965; Polidora, 1967) and depressed brain serotonin concentration in rats (Yuwiler and Louttit, 1961) makes this an intriguing area for further investigation, because many genetic defects of amino acid metabolism result in mental deficiency.

Rats fed excess tyrosine in a low-protein diet develop redness and swelling of the paws, a dark exudate accumulates around the eyes, and histopathological changes occur in several tissues (Lillie, 1932; Heuper and Martin, 1943). The adult rat is more resistant to tyrosine toxicity than the young rat (Schweizer, 1947). Tolerance for tyrosine increases as the protein content of the diet is increased (Harper *et al.,* 1966); but, if the tyrosine content is 10 percent or more, toxicity appears in rats fed an adequate diet (Schweizer, 1947). Addition of ascorbic acid, which is known to be involved in tyrosine metabolism, to a high-tyrosine diet does not prevent signs of tyrosine toxicity (Martin and Hueper, 1943); but threonine, some other amino acids, and thiouracil or cortisol treatment alleviate signs of tyrosine toxicity in rats fed a low-protein diet (Alam *et al.,* 1967).

Signs of tyrosine toxicity are associated with high blood and tissue tyrosine concentrations (Sauberlich, 1961; Alam *et al.,* 1967). Neither *p*-hydroxyphenylpyruvate nor thyroxine causes signs resembling tyrosine toxicity in rats (Boctor, 1967); the syndrome appears to be due to the accumulation of tyrosine itself.

Phenylalanine and tyrosine administered orally in doses of 20 g/day to mental patients apparently did not cause adverse effects (Pollin *et al.,* 1961). A single oral dose of 7 g of phenylalanine given to patients with

Parkinsonism and control individuals was cleared from the blood in 4 h without any adverse reactions (Braham *et al.*, 1969).

Tryptophan

The first amino acid demonstrated to be nutritionally essential (Willcock and Hopkins, 1906), tryptophan was also the first shown to be toxic when fed in excess to the rat (Hicks, 1926). Ingestion of it in excessive amounts (Sauberlich, 1961; Harper *et al.*, 1966) causes growth depression in rats fed a low-protein diet; but, as the protein content of the diet is increased, the rats develop a tolerance for it (Sauberlich, 1961; Harper *et al.*, 1966; Daniel and Waisman, 1968).

Cows given an oral dose of D L-tryptophan (0.7 g/kg of body wt) develop interstitial pulmonary emphysema, apparently due to a product of ruminal metabolism, because injected tryptophan does not have this effect (Carlson *et al.*, 1968). The cow may prove a useful animal model for study of this disease.

Tests involving measurement of urinary excretion of tryptophan metabolites after a tryptophan load indicate that man tolerates well individual doses of as much as 10 g of D L-tryptophan repeated weekly for many weeks (Baker *et al.*, 1964). Adult patients tolerated oral doses of 20–50 mg of tryptophan/kg of body wt (1.5–3.5 g/day /70 kg) without untoward effects; but this amount, administered with a monoamine oxidase inhibitor, produced "intoxication," "drowsiness," and "hyperreflexia" (Oates and Sjoerdsma, 1960). Schizophrenic patients tolerated daily doses of 7–15 g of tryptophan for up to a week; if iproniazid was administered at the same time, however, some quite startling behavioral changes occurred (Pollin *et al.*, 1961). Tryptophan, when administered with an amine oxidase inhibitor, such as iproniazid, can be highly toxic (Hodge *et al.*, 1964). Patients with Parkinsonism tolerated 8–9 g of tryptophan or 100 mg of vitamin B_6 daily for several days without event, but the two administered together caused rapid deterioration in their condition (Hall *et al.*, 1972). However, the administration to 41 patients with depression of 5–7 g of D L-tryptophan daily for 28 days resulted in no adverse effects (Coppen *et al.*, 1967). (The patients were given vitamin B_6; 19 of them also received a monoamine oxidase inhibitor.) Most showed as much improvement in their condition, as did others treated with electroconvulsive therapy.

Certain tryptophan metabolites have been examined as possible bladder carcinogens (Bryan, 1971; Yoshida *et al.*, 1971). There is no direct evidence that tryptophan metabolites produce bladder cancer in man, but animal studies suggest that they may act as cocarcinogens in the presence of specific carcinogens.

Sulfur-containing Amino Acids

In studies of the relative toxicity of ingested amino acids, methionine has consistently proven the most toxic—quantities in excess of 2 percent in diets containing 10 percent of protein cause growth reduction in the rat (Sauberlich, 1961; Benevenga and Harper, 1967). A dietary excess of methionine appears to increase the requirement for pyridoxine. Even a small addition of methionine to a low-protein, pyridoxine-deficient diet reduces the survival of rats (Cerecedo and DeRenzo, 1950).

An excessive intake of methionine causes kidney hypertrophy and marked changes in the pancreatic acinar cells (Earle *et al.*, 1942), as well as severe liver damage. Darkened spleens, caused by marked increases in iron deposition, have occurred routinely in animals fed high-methionine diets (Van Pilsum and Berg, 1950), possibly due to iron accumulation as a result of hemoglobin degradation associated with the anemia that occurs (Klavins and Johansen, 1965). Supplements of glycine and serine (Benevenga and Harper, 1967) afford partial protection against methionine toxicity, apparently by enhancing the rate of oxidation of methionine and thereby reducing its circulating concentration.

Depressed growth characterizes rats ingesting amounts of L-cystine in excess of 2.5 percent in a low-protein diet (Sullivan *et al.*, 1932), in addition to liver damage and kidney necrosis (Curtis and Newburgh, 1927; Earle *et al.*, 1942). Higher amounts cause a greater mortality rate (Graham *et al.*, 1950). The signs of cystine toxicity distinctly differ from those of methionine toxicity, indicating that the latter cannot be attributed to excessive formation or accumulation of cysteine or cystine.

In one study, methionine was administered to adult patients with liver disease in quantities of 8–20 g/day in divided doses for 3–9 days (Phear *et al.*, 1956). In seven patients with portal cirrhosis, neurological deterioration occurred in 1–4 days following total doses of 11–46 g of methionine. When the liver is damaged, large amounts of either protein or amino acids in the diet can cause hepatic coma. However, 10 patients who had not previously exhibited neurological complications tolerated methionine without evidence of neurological change—1 at 20 g/day for 3 days and another at 10–14 g/day for 9 days. Methionine tolerance was impaired in patients with liver disease. Their blood methionine concentrations remained elevated 4 h following intravenous infusion of 6 g of methionine, whereas the blood methionine concentration in three healthy subjects had returned to normal by this time. Administration of chlortetracycline with methionine largely protected susceptible patients from neurological deterioration, suggesting that the intestinal flora may be involved in the toxic effects observed. From a review of earlier work and their own studies, Phear *et al.* (1956) concluded that therapy with methionine was not beneficial, and could be deleterious, to human subjects with liver disease. They did, however, demonstrate that subjects without severe liver involvement could tolerate daily doses of 10–20 g of methionine for short periods without adverse effects. Methionine at 20 g/day for 7 days, but not at 15 g/day, also caused behavioral changes and some gastric distress in schizophrenic patients (Pollin *et al.*, 1961).

Histidine

Rats fed a low-protein diet containing as little as 2 percent of L-histidine suffer severe growth depression, but their tolerance for histidine increases with improvement in either the quality or the quantity of the protein in the diet (Salmon, 1958; Harper *et al.*, 1966). Infant monkeys fed a high-histidine diet (3 g/kg/day) developed a serum hyperlipemia, involving several classes of serum lipids, within 3–4 months (Kerr *et al.*, 1966). The basis for this effect is unknown, but the observation raises the issue of hitherto unrecognized relationships between amino acid and lipid metabolism.

Histidine has not been studied extensively in human subjects, but schizophrenic patients tolerated 20 g/day for a week without evidence of distress (Pollin *et al.*, 1961). In a recent study (Henkin *et al.*, in press), adult volunteers tolerated up to 32 g of histidine/day for short periods; but patients with scleroderma who were given histidine doses of 8–64 g/day for 2-day periods after 2-day intervals experienced a fall in serum–zinc concentration and deterioration of taste acuity.

Threonine

Threonine is one of the least toxic of the amino acids (Sauberlich, 1961). A large intake greatly elevates plasma threonine concentration (Sauberlich, 1961; Alam *et al.*, 1966). Homoserine, which is metabolized essentially like threonine, is also well tolerated by the rat (Cohen *et al.*, 1958). The relatively high tolerance of the rat for threonine seems to be due to the limited effect of an excess of threonine on food intake

(Peng and Harper, 1970) and to the ease of oxidation of threonine metabolites.

Lysine

Chicks (Anderson *et al.*, 1951) or rats (Jones *et al.*, 1966) experience retarded growth when lysine comprises about 5 percent of a diet marginal-to-low in protein. The effect of excess lysine is moderate relative to that of methionine or tryptophan (Sauberlich, 1961). Large amounts of lysine (10–25 percent of the diet) fed to pregnant rats from the 5th to 15th day of gestation increased litter mortality, although surviving fetuses suffered no gross malformations (Cohlan and Stone, 1961). Lysine accumulates in plasma proportionately to the dietary excess (Zimmerman and Scott, 1965).

Lysine·HCl has been used as an adjuvant to mercurial diuretics in treating refractory edema (Lasser *et al.*, 1960). When 10–40 g/day were administered for up to 6 days orally in 4 doses to patients with congestive heart failure, occasionally abdominal cramps or diarrhea occurred when the dose was high but disappeared when it was reduced. Oral administration of 300 mg of lysine ·HCl/kg of body wt to a normal child produced no noticeable effects, but led to coma in a child with lysinemia, a genetic defect of lysine metabolism (Colombo *et al.*, 1967).

Arginine

Rats fed low-protein diets tolerate a high intake of arginine better than high intakes of most of the essential amino acids (Sauberlich, 1961). Four percent of L-arginine, fed in a low-protein diet, depresses their growth (Harper *et al.*, 1966), but the same amount fed in a diet containing an adequate amount of protein causes no adverse effect (Jones *et al.*, 1966). Although the tolerance of the rat for arginine is greater when the protein content of the diet is increased or if the quality of a low-protein diet is improved (Harper *et al.*, 1966), the growth of rats consuming an adequate amount of protein is depressed if arginine content is increased sufficiently (Schimke, 1963).

Arginine has been tested, with variable results, as a therapeutic agent for controlling blood ammonia concentration in man. Najarian and Harper (1956) infused 15 patients who had elevated blood ammonia with 25–50 g of arginine·HCl per day and observed improved mental status and reduced blood ammonia concentrations. Fahey *et al.* (1957) infused 29–39 g of arginine·HCl over a period of 60–120 min into patients with liver disease without observing beneficial effects. In

another study (Reynolds *et al.*, 1958), patients with hepatic insufficiency were infused intravenously with 30 g of arginine·HC1 daily without evidence of side effects or much beneficial effect.

Arginine·HC1 has also been used as an adjuvant to diuretics for patients with edema (Ogden *et al.*, 1961). Intravenous administration of
up to 42 g of arginine·HC1 daily for as long as 9 days proved to be
"a safe and effective method of producing hyperchloremic acidosis and
restoring responsiveness to mercurial diuretics in patients" with fluid
retention. Single infusions of 30 g have also been shown to induce insulin response in healthy women and to be well tolerated (Merimee *et al.*,
1965). Amounts up to 25 g/day for 10 days administered to 9–12-year-
old children with cystic fibrosis (Solomons *et al.*, 1971) appeared to improve fat absorption without producing observable side effects.

Glycine and Serine

The degree of toxicity or growth depression due to an excessive intake
of glycine depends upon the amino acid composition and the protein
content of the diet. When the glycine content of the diet is 5 percent
or less, the protein content is 10–28 percent, and the vitamin content
adequate, little adverse effect is observed (Naber *et al.*, 1956; Benevenga
and Harper, 1967). Certain vitamin deficiencies decrease the tolerance
of animals for glycine (Harper *et al.*, 1970). Extensive clinical and nutritional studies in which human subjects consumed large amounts of glycine (Harper, 1973) indicate that man's tolerance for this amino acid is
high.

As much as 5 percent of D L–serine added to the diet of a chick or rat
has not resulted in severe growth reduction or death (Graham *et al.*,
1950; Naber *et al.*, 1956; Sauberlich, 1961; Benevenga and Harper,
1967), but inclusion of 4–6 percent of L-serine in a 10 percent casein
diet depressed the growth of rats by about 25 percent (Benevenga and
Harper, 1967).

Other Dispensable Amino Acids

The dispensable amino acids are tolerated in excess by animals much
better than are the essential ones, only moderate growth depressions
occurring with substantial excesses of many of the former. No unique
toxic effects of the L-forms of the dispensable amino acids have been
reported. Glutamic acid is well tolerated by the rat in amounts as high
as 5 percent, and probably up to 10 percent, in diets with low-to-mod-
erate protein content (Sauberlich, 1961; Hepburn and Bradley, 1964).

Glutamine in the amount of 6 percent in an adequate amino acid diet exerted no adverse effect (Hepburn and Bradley, 1968). Alanine exerted no adverse effect on rats fed 5 percent in a low-protein diet (Sauberlich, 1961). The growth of rats fed 5 percent of L-aspartic acid or L-asparagine in a low-protein diet was depressed by 50 percent (Sauberlich, 1961), but no adverse effect was observed with rats fed a diet adequate in protein.

Many food proteins contain from 20 to 35 percent of glutamic acid. Thus, adults in the United States commonly consume 20–35 g/day of glutamic acid throughout life, with children consuming proportionately less, depending on their energy and protein intake. These quantities in the food are obviously innocuous. Diets used by Nakagawa *et al.* (1964) in studies of the amino acid requirements of boys 10–12 years of age provided 12.75 g/day or more of free glutamic acid.

Concern about possible adverse effects from glutamate ingestion has arisen from observations that subcutaneous injection of 4–8 g/kg of sodium L-glutamate into neonatal mice produced changes in the retina (Lucas and Newhouse, 1957; Anonymous, 1970) and that large parenteral or oral doses (9.5–4 g/kg of body wt) will produce neuropathologic, growth, and endocrine changes in neonatal mice (Olney 1969; Olney and Sharpe, 1969; Anonymous, 1970). Large doses are required to produce these lesions, and even larger ones in older animals.

Although large doses of monosodium glutamate administered to neonatal animals that have a slowly developing nervous system will produce retinal and neuropathological lesions, other species with more fully developed nervous systems are more resistant (Anonymous, 1970); observations on both children and adult man indicate that human tolerance for this amino acid is high.

D-Amino Acids

Berg (1953, 1959) has reviewed information about the nutritional value and metabolism of the D-amino acids. His and subsequent information indicates, with the exceptions of D-aspartic acid and D-alanine, they are tolerated better than the L-forms (Harper *et al.*, 1970).

AMINO ACID ANTAGONISMS

Branched-chain Amino Acids

Excessive amounts of the branched-chain amino acids depress the growth of rats (Russell *et al.*, 1952). The growth depression caused by excess leucine in low-protein diets can be largely overcome by the concomitant addition of isoleucine and valine (Benton *et al.*, 1956b). In view of the structural similarity of these molecules, the growth depression caused by leucine has been attributed to an amino acid antagonism. Excess isoleucine or valine caused little growth depression; but, when leucine was growth-limiting in a diet containing a mixture of amino acids, growth was improved by omission of valine from the mixture (Benton *et al.*, 1956b). It thus appears that an excessive amount of any one branched-chain amino acid can increase the requirement for the other two.

The rat adapts to an excessive intake of leucine, the time required to do so depending upon the amount of leucine in the diet (Spolter and Harper, 1961). Moreover, leucine in considerable excess is tolerated well by rats fed a diet containing an adequate amount of protein (Sauberlich, 1961). Increased litter mortality and a decrease in the weight of surviving fetuses occurred when pregnant rats were fed casein diets containing 25 percent of leucine by weight, but no gross malformations appeared among the surviving litters (Cohlan and Stone, 1961; Persaud, 1969).

In studies of the amino acid requirements of adults, Rose (1957) and associates and Linkswiler *et al.* (1960) observed that subjects consuming isoleucine-deficient diets experience more severe nausea and anorexia than those consuming diets deficient in other amino acids. The leucine content of the basal amino acid mixture was high, raising the possibility that branched-chain amino acid antagonism may also occur in man. However, if analogy to the rat studies is valid, this antagonism would occur only if the diet were low in isoleucine and valine.

Leucine, either orally or parenterally, causes a severe hypoglycemic reaction in persons who are genetically susceptible (DiGeorge and Auerbach, 1960). This amino acid stimulates insulin production. However, 10 g of leucine administered orally to normal adults is without effect (Fajans *et al.*, 1963), and 14–20 g causes only a small depression in blood glucose concentration (Fajans *et al.*, 1963; Gopalan, 1968).

Observations by Gopalan (1970) and associates indicate that the high leucine content of certain low-protein cereal diets may contribute to the development of pellagra. They also observed that oral administration of leucine (10 g/day for 5 days) to pellagrins or to normal subjects depressed the ability of erythrocytes to synthesize pyridine nucleotides and that inclusion of leucine in a nonpellagragenic diet for pups converted it to one that caused pellagra. Further reduction of the leucine content of a maize diet prevented pellagra in pups (Belavady, 1967; Gopalan, 1970). Hankes *et al.* (1971), in a study of pellagra in South Africa, also concluded that the high leucine content of maize depressed synthesis of nicotinamide adenine dinucleotide phosphate in the body.

Lysine–Arginine Antagonism

Growth of chicks fed a diet containing an adequate amount of arginine was depressed by a supplement of 2 percent of lysine, but optimal growth was restored by further addition of arginine (Jones, 1964). The growth-depressing effect of excessive dietary lysine is more severe in a strain of checks selected for a high arginine requirement than in one selected for a low requirement. The difference in arginine requirement (Nesheim, 1968a) is exaggerated when casein, which is particularly rich in lysine, is the dietary protein.

Demonstration of a lysine–arginine antagonism in mammals is complicated by their ability to synthesize arginine, but has been demonstrated in the guinea pig (O'Dell and Regan, 1962) and in the rat (Jones *et al.,* 1966). Lysine–arginine antagonism appears to involve increased loss of arginine, probably as a result of competition for reabsorption in the kidney (Nesheim, 1968b) and induction of arginase by lysine, resulting in increased arginine degradation.

AMINO ACID EXCESS IN MAN

Snyderman *et al.* (1968) administered to infants single loads of most of the individual amino acids in amounts equivalent to 5 or more times the requirement. Preliminary reports of these studies, which have not been reported in detail, do not mention adverse effects. Plasma amino acid patterns have been altered such that the amino acid administered is greatly elevated; while certain other amino acids, depending upon the one administered, are depressed. Chronic loading for 1-week periods provided evidence that the body adapted to a high leucine intake and, after the adaptation period, cleared leucine from the blood rapidly (Holt, 1967).

Information about effects of accumulations of individual amino acids in human subjects has also been obtained from study of genetic defects of amino acid metabolism in which the initial enzyme for degradation of an amino acid is lacking. The amino acid for which the catabolic enzyme is missing accumulates in body fluids, even when the amount consumed is not excessive. Observers have described defects involving all the amino acids except threonine (Hsia, 1966; Nyhan, 1967; Milne, 1968; Stanbury *et al.,* 1972); they commonly cause mental deficiency and neurologic disorders. Probably the best known and most common genetic defect of this type is phenylketonuria, in which the absence of phenylalanine hydroxylase from the liver results in intolerance of phenylalanine. This disease leads to mental deficiency, decreased pigmentation, diminished serotonin formation, and various other changes, all of which are associated with high phenylalanine concentrations in body fluids. Treatment with a low-phenylalanine diet results in improved growth and development and intelligence (Berman et al., 1966; Berry et al., 1967; Kennedy et al., 1967). This is, in essence, nutritional control of an endogenous amino acid toxicity.

AMINO ACID IMBALANCE

The concept of amino acid imbalance arose from observations that supplements of tryptophan-deficient proteins depressed the growth of rats fed a niacin-deficient, low-protein diet (Krehl et al., 1945; Harper et al., 1970). After the discovery that amino acid imbalances could be demonstrated with animals fed on diets containing nicotinic acid (Harper, 1958a; Salmon, 1958), emphasis shifted away from consideration of amino acid imbalance as a problem related specifically to nicotinic acid–tryptophan deficiency and toward amino acid imbalance as a more general condition involving nutritional and metabolic interrelationships.

Research has provided many examples of growth depressions due to amino acid imbalances. Growth of rats fed a rice diet increases with increasing increments of threonine until, with 0.3 percent of D L-threonine, imbalance occurs, at which point growth rate is depressed. Additional lysine is then required to correct the imbalance and prevent the growth depression (Rosenberg et al., 1959). Growth of rats fed a low-protein diet is depressed by inclusion of an amino acid mixture devoid of threonine, and the imbalance so created is corrected by a supplement of threonine (Harper, 1958a). The growth depression caused by imbalances of both types becomes less severe as the protein content of the diet is increased toward adequacy. This is not surprising, as the amount of limiting amino acid in a diet is increased when the protein content is increased; and, hence, the degree of imbalance is less in relation to the total protein content of the diet.

Effects of amino acids on fat deposition in the liver of rats fed a low-protein diet containing choline have been attributed to amino acid imbalances. For example, the fat content of the livers of rats fed a 9 percent casein diet is nearly normal, but the fat content of the livers of rats fed the same diet supplemented with methionine is elevated. This elevation can be largely prevented by further supplementation with threonine (Harper, 1958b).

The alteration in amino acid pattern that results in accumulation of fat in the liver differs distinctly from that leading to growth depressions attributable to amino acid imbalances. The supplement causing liver fat accumulation consists of a quantity of the growth-limiting amino acid; the supplement causing growth depression consists of a quantity of an amino acid other than the one that is growth limiting. Since restriction of caloric intake, without concomitant restriction of protein intake, alleviates the fatty infiltration of the liver without significantly retarding growth (Yoshida *et al.,* 1961), the effect is apparently the result of protein–calorie imbalance rather than amino acid imbalance, even though it is brought about by a change in the amino acid balance of the diet.

The earliest investigations of amino acid imbalances showed that the growth depressions observed were associated with depressions of food intake. The low intake of food was assumed to be due to a metabolic derangement; but, when it was discovered subsequently that food intake could be depressed within 4 to 8 h by a dietary imbalance of amino acids, the basic problem appeared to involve food intake regulation directly (Harper, 1967). To prevent the depressions of growth and food intake, it was necessary to increase the content of the limiting amino acid in the diet. Nevertheless, studies in which food intake of animals fed diets with imbalances that have been equalized with those of controls point to the conclusion that imbalances do not affect efficiency of utilization of the limiting amino acid under these conditions; the extra quantity of the limiting amino acid may be required only to stimulate food intake and may subsequently be degraded (Harper *et al.,* 1970).

The growth and food intake depressions caused by amino acid imbalances in diets that are not deficient in nicotinic acid are most severe during the first few days. Thereafter, food intake increases gradually and, depending upon the degree of imbalance, growth rate may eventually approach that of the control group. The time required for adaptation of rats to diets in which imbalances have been created also depends upon the severity of the imbalance (Leung *et al.,* 1968a). It is shortened by conditions that tend to stimulate food intake.

Dietary amino acid balance can also influence food preference. In general, the rat will select a protein-free diet or a diet with a balanced amino acid pattern over one in which there is an amino acid imbalance (Harper, 1967; Leung *et al.,* 1968b). If one assumes that the basis for these food preferences is rejection by the animal of a diet with an amino acid imbalance, the rejection is not on the basis of nutritional inadequacy, as the protein-free diet that is completely inadequate is selected over the more adequate diet with an amino acid imbalance.

The most reproducible and most extensively studied biochemical change that occurs in animals fed a diet with an amino acid imbalance is a greatly altered plasma amino acid pattern following ingestion of a meal. In rats fed a diet with an imbalance, the plasma concentration of the growth-limiting amino acid falls rapidly; and the concentrations of most of the amino acids added to create the imbalance rise substantially, such that the plasma amino acid pattern resembles very closely that of an animal fed a diet that is severely deficient in an amino acid. The pattern in muscle resembles that in plasma. In brain, the concentration of the growth-limiting amino acid is depressed, although the concentrations of other amino acids in brain are not greatly elevated (Peng *et al.*, 1972).

The time of occurrence of the plasma and brain amino acid changes corresponds closely to the time of occurrence of food intake depression. Also, consumption of a protein-free diet, or any of the various balanced diets the rat will select in preference to an imbalanced diet, results in restoration of a balanced plasma amino acid pattern. All of the observations taken together indicate that a close association exists between the alterations in plasma amino acid pattern and changes in food intake and preference.

Accumulations of essential amino acids in blood and body fluids, such as are observed after ingestion of deficient, imbalanced, and high-protein diets and even after ingestion of a diet containing a large excess of one amino acid, might indicate that protein-synthesizing sites are saturated and thereby provide a general signal for curtailment of further amino acid ingestion. In animals fed deficient or imbalanced diets, the low concentration of the limiting amino acid could serve as a signal that leads to depression of intake of a diet that, if ingested in large amount, might lead to pathologic consequences, as has been shown with rats force-fed diets devoid of one amino acid. Leung and Rogers (1969) have identified a site in the brain that is responsive to a low concentration of an individual amino acid.

Distinguishing between effects of amino acid deficiencies and amino acid imbalances in human subjects consuming natural diets is difficult except by elaborate and well-controlled experiments. However, in some studies in which amino acid supplements were tested for their ability to improve low-protein diets consisting largely of cereal products, amino acid imbalances may have been created. Certain amino acid additions decreased nitrogen retention in adult men consuming a rice diet (Hundley *et al.*, 1957); and supplement of methionine depressed nitrogen retention of children consuming a vegetable protein diet based on maize; this was prevented by a further supplement of isoleucine (Scrimshaw *et al.*, 1958). The depressed nitrogen retention was associated with anorexia (G. Arroyave, personal communication).

PHYSIOLOGICAL AND METABOLIC RESPONSES

Two common effects of ingestion of diets with different types of disproportions of amino acids are: Food intake is depressed, even though the depressions may differ in severity and duration depending upon the type and degree of disproportion; and plasma amino acid concentrations are elevated by all of these conditions, even though the plasma amino acid patterns may differ distinctly. If the information on this subject is fitted into the framework of the concept of homeostasis, all of the effects of ingestion of diets containing disproportionate amounts of amino acids can be interpreted as responses that occur as a result of exceeding the capacity of the organism to maintain homeostasis of plasma amino acid concentrations. To emphasize that depressed growth and food intake are detrimental rather than to emphasize the survival value of a reduced intake of a diet that can produce adverse effects and the importance of food intake regulation as a mechanism contributing to homeostasis of plasma amino acid concentrations is, perhaps, misleading.

The amino acid-degrading capacity of animals, particularly their capacity to degrade the essential amino acids, is highly responsive to change in protein intake (Ashida and Harper, 1961; Muramatsu and Ashida, 1962; Ashida, 1963; Knox and Greengard, 1965). The activities of enzymes of amino acid catabolism tend to be low in animals fed a low-protein diet, thereby limiting their capacity to remove amino acids ingested in excess. Under these conditions, ingestion of a high-protein diet, a diet with an amino acid imbalance, or a diet with a surplus of one or more amino acids will cause amino acids to accumulate in body fluids. If this is the result of a high-protein intake, with the dietary amino acids in a reasonably well-balanced pattern, a part of the excess can be used to synthesize tissue proteins. Moreover, in animals with a substantial intake of high-quality protein, most of the enzymes of amino acid catabolism increase greatly in activity within a short time; and thus the capacity to degrade a surplus of amino acids increases rapidly and homeostasis is restored, but at a higher level of protein metabolism.

If a surplus of amino acids arises from ingestion of a diet with an amino acid imbalance, or one containing a large excess of one amino acid, the amount of balanced protein ingested remains low. The amino acids that are in surplus cannot be used to synthesize tissue proteins; and amino acid-degrading enzymes do not undergo adaptation, either rapidly or extensively, as they do in animals fed a high-protein diet, presumably because conservation of a limited supply of amino acids in a well-balanced pattern has evolved as a mechanism contributing to the survival of the animal receiving an inadequate amount of protein. If amino acid-degrading capacity is exceeded by continuous ingestion of such diets, amino acids will continue to accumulate in body fluids; and, unless some further regulatory mechanism comes into play, the accumulation may be so great as to be toxic. This occurs with animals that are force-fed diets devoid of one amino acid (Sidransky and Farber, 1958), the most severe form of imbalance, and with human subjects who have inborn errors of amino acid metabolism.

Because large amounts of amino acids spill into the urine only after toxic levels occur in tissues; and because intestinal absorption of amino acids is not curtailed even when the diet consists almost exclusively of protein, the sole remaining way to prevent accumulation of amino acids is to reduce the influx through depression of food intake.

From this viewpoint depression of food intake must be looked upon as a normal homeostatic response that has survival value, even though it may be considered an adverse effect from the viewpoint that rapid growth is desirable. Evidence for its survival value is obtained from studies in which stimulation, through cold exposure, of the food intake of rats fed a high-methionine diet was shown to shorten their survival (Beaton *et al.*, 1963). Force-feeding rats a diet devoid of a given amino acid (Sidransky and Farber, 1958) has the same effect. Food intake regulation can thus be viewed as a homeostatic mechanism in which the entire organism functions as a feedback system contributing to the regulation of free amino acid concentrations of plasma and tissues by reducing the influx of amino acids when plasma concentrations are excessive (Harper, 1968).

Severe adverse effects from ingestion of diets containing disproportionate amounts of amino acids presumably occur only when the capacity of the homeostatic mechanisms regulating blood and tissue amino acid concentrations is exceeded. This occurs most readily in animals fed low-protein diets containing an excessive amount of an individual amino acid, for such animals have a limited capacity for growth and show low activities of amino acid-degrading enzymes; they must therefore depend upon depressed food intake to limit the accumulation in body fluids of the amino acid that is in excess in the diet.

The capacity of homeostatic mechanisms is limited; the point at which food intake drops is presumably the resultant of the action of the mechanisms that stimulate the organism to eat until its energy requirement is met and those that inhibit its intake of a diet that will drastically alter its *milieu intérieur*. If the compromise level of food intake still results in tissue accumulation of the amino acid in excess in the diet, signs of toxicity will develop; in any event if food intake is depressed below the amount required to provide the maintenance levels of energy and other nutrients, deterioration will occur. Whether the deterioration will continue until death ensues depends upon a second aspect of homeostasis—that is, adaptation. The animal fed a low-protein diet is at a disadvantage here too, for the activity of amino acid-degrading enzymes tends to increase as an animal matures and as protein intake increases. Nevertheless, even with intakes of amino acids that cause signs of toxicity, adaptation may occur with time, signs of toxicity abate, and food intake and growth rate improve—presumably because the capacity for amino acid degradation gradually increases, provided food intake has not been too severely depressed. The end result is not necessarily prevention of abnormality of the amino acid patterns of blood and body fluids; it may only be prevention of accumulations of amino acids that are great enough to cause adverse effects.

IMPLICATIONS FOR HUMAN NUTRITION

One of the underlying assumptions of methods for evaluating the nutritional quality of proteins is that amino acids present in the diet in excess of the amounts that can be used for protein synthesis are innocuous (Oser, 1945; Block and Mitchell, 1946–47). The direct relationship observed between chemical scores calculated from the amino acid composition of proteins and estimates of protein quality obtained using biological methods for a variety of proteins over a fairly wide range of intakes lends support to this assumption, indicating that disproportionate amounts of amino acids encountered naturally do not commonly result in adverse effects. Adverse effects observed experimentally are usually the result of amino acid disproportions greater than those encountered in nature and greater than could be created through a rational program of fortification of foodstuffs with amino acids.

Table 1 provides information about the tolerance of the rat for amino acids and about quantities of amino acids that might be consumed in foods by men. The first column gives a rough estimate of the percentage of each of the amino acids that has been observed to cause adverse effects in the young, growing rat fed a low-protein diet. The essential, and below them the dispensable, amino acids are listed in order of increasing tolerance. Leucine has been listed lower in the sequence than would appear appropriate, because tolerance for it increases greatly as

TABLE 1 Amino Acid Requirements of the Rat and Man, Tolerance of the Rat for Excessive Intakes of Individual Amino Acids, and Amounts of Amino Acids in Wheat and Meat Diets

Amino Acid	Level Causing Adverse Effects in the Rat[a] (% diet)	Requirements for Young Rat[b] (% diet)	In 70% Casein Diet[c] (% diet)	Requirements for Adult Human[d]		Grams in 2,800 kcal of			Requirements ± for 12-kg Child[h], 1–2 yr (g/day)	Grams in 1,200 kcal of	
				70-kg Male (g/day)	58-kg Female (g/day)	Whole Wheat[e]	Mixed Diet[f]	Round Steak[g]		Whole Wheat[e]	Round Steak[g]
Methionine	2	0.5	2.3	0.55	0.46	1.8	2.9	11.4	0.41	0.73	4.6
Tryptophan	1.5	0.12	1.0	0.16	0.13	1.5	1.5	5.4	0.17	0.58	2.2
Histidine	2	0.26	2.1	–	–	2.4	2.5	13.4	0.3	0.95	5.3
Tyrosine	3	–	4.5	–	–	4.4	5.1	13.0	–	1.7	5.2
Cystine	3	–	0.3	–	–	2.6	1.7	5.8	–	1.0	2.3
Phenylalanine	4	0.9	4.1	0.85	0.70	5.8	5.3	15.8	1.2	2.3	6.3
Leucine	2.5	0.7	7.0	0.88	0.72	8.0	9.2	37.8	1.2	3.1	15.2
Isoleucine	5	0.55	4.7	0.66	0.55	5.3	6.3	24.2	0.78	2.0	9.5
Valine	5	0.55	5.2	0.75	0.62	5.5	6.7	21.4	0.84	2.2	8.5
Lysine	5	0.9	5.8	0.66	0.55	3.2	8.6	40.4	0.96	1.3	16.2
Threonine	5	0.5	3.2	0.46	0.38	2.4	5.2	20.4	0.56	1.2	8.2

Serine	4	—	4.6	—	5.3	—	16.1	—	2.2	6.4
Arginine	4	0.2	2.8	—	5.7	—	24.6	—	2.2	9.7
Glycine	4–5	—	1.4	—	7.3	—	23.7	—	2.8	9.4
Aspartic acid	5	—	5.2	—	6.5	—	35.8	—	2.5	14.3
Proline	5	—	8.3	—	12.4	—	19.0	—	4.8	7.5
Alanine	>5	—	2.3	—	4.2	—	22.2	—	1.6	8.8
Glutamic acid	7	—	16.0	—	32.2	—	58.0	—	14.5	23.2
Protein	12–15	—	40	32	118	122	464	12	47	182

aBased on information from studies on rats fed a low-protein diet compiled in Harper et al., 1970.

bRama Rao et al., 1960.

cOrr and Watt, 1957 (assuming 16 percent N in casein).

dBased on Table 10, Chapter 2.

eAssuming 330 kcal/100 g and 14 percent protein (Orr and Watt, 1957; Watt and Merrill, 1963).

fFrom FNB, 1959.

gOrr and Watt, 1957 (assuming broiled, lean beef = 189 kcal/100 g and 31.3 percent protein).

hOrr and Watt, 1957 (assuming a requirement of 1 g protein/kg/day of the amino acid composition of cows' milk).

the protein content of the diet is increased and because the rat adapts rapidly to a diet containing an excess of it.

Most adverse effects of essential amino acids have been demonstrated in young animals fed a low-protein diet. The young animal is much more susceptible to adverse effects from an amino acid load than is the mature animal. Also, the animal fed a low-protein diet is more suscep- tible than one fed an adequate amount of protein; the animal fed a diet deficient in certain vitamins, especially nicotinic acid, pyridoxine, and vitamin B_{12}, is more susceptible than one receiving adequate amounts of all vitamins. Furthermore, adverse effects from excesses of individual amino acids usually become less severe with time unless the load is so great that a distinctly toxic reaction develops, suggesting that animals undergo various types of adaptations that increase their ability to tolerate amino acids in excess.

Examination of Table 1 indicates that the essential amino acids are generally less well tolerated than the dispensable ones, except for tyro- sine and cystine, and that amino acids entering into a variety of meta- bolic pathways and serving as precursors of a variety of biologically active substances tend to be less well tolerated than are those with less- complex metabolic interactions. Methionine is the least well tolerated of the nutritionally important amino acids, but even the young rat fed a low-protein diet will tolerate an amount equal to 3 times the require- ment. The tolerance of tryptophan is about the same on a weight basis; but this is about 12 times the requirement, and the effects of an excess of tryptophan tend to be less severe than those of an excess of methio- nine. Amounts of other amino acids ranging from four to tenfold the amounts required for rapid growth cause growth depressions; yet an 80 percent casein diet, which causes only a transitory retardation of growth (Harper, 1965), contains amounts of seven of the essential amino acids in excess of those observed to depress severely the growth of rats fed a low-protein diet.

The amino acid requirements of man and the amounts of amino acids consumed when energy needs are met from different food sources are also listed in Table 1. An individual existing exclusively on a meat diet would be ingesting many times the requirements for most amino acids in a well-balanced pattern, and he can do so without adverse effects (McClellan and Dubois, 1930). Healthy male volunteers who consumed a purified diet in which 70 percent of the energy was from protein also showed no obvious ill effects (Calloway and Margen, 1968). Thus it is evident that normal man, like laboratory animals, can tolerate large in- takes of amino acids in a well-balanced pattern. Such large intakes of amino acids cause liver and kidney hypertrophy in animals (Addis *et al.*, 1926, 1940) and evidently shorten their life span (Ross, 1959, 1961), suggesting that such a metabolic load does place a strain on regulatory mechanisms even though specific signs are not evident.

In human subjects receiving single doses of individual essential amino acids, either orally or intravenously, in amounts that exceed the requirement by tenfold or more, the major adverse effect is mild gastric distress. This was observed mainly when the amino acid was administered in a single large dose on an empty stomach. The dispensable amino acids arginine and glycine are well tolerated in large doses, as is glutamic acid (except in the case of some individuals who are susceptible to a specific transitory syndrome produced by this amino acid when it is ingested, essentially without food, in amounts of 3 g or more on an empty stomach). Evidence indicates that doses of methionine, isoleucine, and threonine in excess of tenfold the requirement, administered intravenously, may cause nausea, febrile reactions, or headaches. Evidence also exists that methionine, and probably other amino acids, should not be administered to patients with liver dysfunction in doses of tenfold the requirement, that tryptophan in comparable amounts can produce a flushing reaction, and that tryptophan and methionine administered in doses of this order to mental patients treated with monoamine oxidase inhibitors can produce disorientation. Few amino acids, with the exception of glycine, have been administered individually to human subjects in long-term studies, but the ordinary mixed diet of adults in the United States provides many amino acids in 5–10 times the required amounts.

Apart from the observations that leucine may be high enough in maize and related cereal grains to contribute to the development of pellagra when tryptophan and niacin intakes are low, little evidence suggests that individual amino acids are present in foodstuffs in quantities large enough to cause adverse effects. Also, quantities of individual amino acids great enough to cause untoward reactions could not rationally be used as supplements to foodstuffs. The amounts of amino acids (2–5 percent of the diet) found to cause adverse effects in the rat fed a low-protein diet would represent intakes of 10–25 g/day by an adult man consuming about 500 g of dry matter and 2,800 kcal/day. Even general fortification of the diet with specific amino acids would be unlikely to increase the intake of individual amino acids by more than 1–2 g/day; and, of course, these should be only the amino acids that are in short supply in the diet.

Only if individual amino acids were administered regularly as pharmaceutic or therapeutic agents, or for some other purpose, would there be much likelihood of approaching the amounts shown to cause adverse effects in animal studies or in such human trials as have been conducted. Even then, unless protein intake or total food intake was low or the subject was otherwise debilitated, amounts of amino acids administered orally would have to be very large indeed to cause more than mild adverse effects.

The extent to which an amino acid imbalance can be created in human diets through the use of amino acid supplements is not clear; but this possibility deserves consideration as suggested in two experimental trials. An amino acid supplement is of value only if the diet is primarily deficient in that amino acid. A supplement of an amino acid other than the one that is most deficient in the diet is at best innocuous and may, if the analogy to animal experiments is valid, depress food intake. This is likely to occur only if the diet is low in protein and marginal in some essential amino acids. Although such effects may not be serious, in view of the ability of animals to adapt to diets with amino acid imbalances, they are certainly not desirable and can be avoided readily by ensuring that any supplement provided makes the diet complete in all respects.

REFERENCES

Addis, T., L. L. MacKay, and E. M. McKay. 1926. The effect on the kidney of the long continued administration of diets containing an excess of certain food elements. I. Excess of protein and cystine. J. Biol. Chem. 71:139–166.

Addis, T., D. D. Lee, W. Lew, and D. J. Foo. 1940. Protein content of the organs and tissues at different levels of protein consumption. J. Nutr. 19:199–205.

Alam, S. Q., R. V. Becker, W. P. Stucki, Q. R. Rogers, and A. E. Harper. 1966. Effect of threonine on the toxicity of excess tyrosine and cataract formation in the rat. J. Nutr. 89:91–96.

Alam, S. Q., A. M. Boctor, Q. R. Rogers, and A. E. Harper. 1967. Some effects of amino acids and cortisol on tyrosine toxicity in the rat. J. Nutr. 93:317–323.

Anderson, J. O., G. F. Combs, A. C. Groschke, and G. M. Briggs. 1951. Effect on chick growth of amino acid imbalances in diets containing low and adequate levels of niacin and pyridoxine. J. Nutr. 45:345–360.

Anonymous. 1970. Monosodium glutamate—studies on its possible effects on the central nervous system. Nutr. Rev. 28:124–129.

Ashida, K. 1963. Diets and tissue enzymes. Pages 159–184 *in* A. A. Albanese, ed. Newer methods of nutritional biochemistry. Vol. I. Academic Press, New York.

Ashida, K., and A. E. Harper. 1961. Metabolic adaptations in higher animals. VI. Liver arginase activity during adaptation to high protein diet. Proc. Soc. Exp. Biol. Med. 107:151–156.

Auerbach, V. H., H. A. Waisman, and L. B. Wyckoff, Jr. 1958. Phenylketonuria in the rat associated with decreased temporal discrimination learning. Nature 182: 871–872.

Baker, E. M., J. E. Canham, W. T. Nunes, H. E. Sauberlich, and M. E. McDowell. 1964. Vitamin B_6 requirement for adult man. Am. J. Clin. Nutr. 15:59–66.

Beaton, J. R., T. A. Orme, A. Turner, and J. Laufer. 1963. Metabolic effects of dietary protein level in cold-exposed rats. Can. J. Biochem. Physiol. 41:139–147.

Belavady, B., T. V. Madhavan, and C. Gopalan. 1967. Production of nicotinic acid deficiency (black-tongue) in pups fed diets supplemented with leucine. Gastroenterology 53:749–753.

Benevenga, N. J., and A. E. Harper. 1967. Alleviation of methionine and homocystine toxicity in the rat. J. Nutr. 93:44–52.

Benton, D. A., A. E. Harper, H. E. Spivey, and C. A. Elvehjem. 1956a. Phenylalanine as an amino acid antagonist for the rat. Arch. Biochem. Biophys. 60:156–163.

Benton, D. A., A. E. Harper, H. E. Spivey, and C. A. Elvehjem. 1956b. Leucine, isoleucine and valine relationships in the rat. Arch. Biochem. Biophys. 60:147–155.

Berg, C. P. 1953. Physiology of D-amino acids. Physiol. Rev. 33:145–189.

Berg, C. P. 1959. Utilization of D-amino acids. Pages 57–96 in A. A. Albanese, ed. Protein and amino acid nutrition. Academic Press, New York.

Berman, P. W., H. A. Waisman, and F. N. Graham. 1966. Intelligence in treated phenylketonuric children; a developmental study. Child Dev. 37:731–747.

Berry, H. N., B. S. Sutherland, B. Umbarger, and D. O'Grady. 1967. Treatment of phenylketonuria. Am. J. Dis. Child. 113:2–5.

Block, R. J., and H. H. Mitchell. 1946–47. The correlation of the amino acid composition of proteins with their nutritive value. Nutr. Abstr. Rev. 16:249–278.

Boctor, A. M. 1967. Some nutritional and biochemical aspects of tyrosine toxicity and lysine availability. Ph.D. Thesis. Massachusetts Institute of Technology, Cambridge.

Braham, J., I. Sorova Pinhas, M. Crispin, R. Golan, N. Levin, and A. Szeinberg. 1969. Oral phenylalanine and tyrosine tolerance tests in Parkinsonian patients. Br. Med. J. 2:552–554.

Bryan, G. T. 1971. The role of urinary tryptophan metabolites in the etiology of bladder cancer. Am. J. Clin. Nutr. 24:841–847.

Burrill, L. M., and C. Schuck. 1964. Phenylalanine requirements with different levels of tyrosine. J. Nutr. 83:202–208.

Calloway, D. H., and S. Margen. 1968. Human response to diets very high in protein. Fed. Proc. 27:725. (A)

Carlson, J. R., I. A. Dyer, and R. J. Johnson. 1968. Tryptophan-induced interstitial pulmonary emphysema in cattle. Am. J. Vet. Res. 29:1983–1989.

Cerecedo, L. R., and E. C. DeRenzo. 1950. Protein intake and vitamin B_6 deficiency in the rat. III. The effect of supplementing a low-protein, vitamin B_6 deficient diet with tryptophan and with other sulfur-free amino acids. Arch. Biochem. 29:273–280.

Cohen, H. P., H. C. Choitz, and C. P. Berg. 1958. Response of rats to diets high in methionine and related compounds. J. Nutr. 64:555–569.

Cohlan, S. Q., and S. M. Stone. 1961. Effects of dietary and intraperitoneal excess of L-lysine and L-leucine on rat pregnancy and offspring. J. Nutr. 74:93–95.

Colombo, J. P., W. Bürgi, R. Richterich, and E. Rossi. 1967. Congenital lysine intolerance with periodic ammonia intoxication; a defect in L-lysine degradation. Metabolism 16:910–925.

Coppen, A., D. M. Shaw, B. Herzberg, and R. Maggs. 1967. Tryptophan in the treatment of depression. Lancet 2:1178–1180.

Curtis, A. C., and L. H. Newburgh. 1927. The toxic action of cystine on the liver of the albino rat. Arch. Intern. Med. 39:828–832.

Daniel, R. G., and H. A. Waisman. 1968. The effects of excess amino acids on the growth of the young rat. Growth 32:255–265.

DiGeorge, A. M., and V. H. Auerbach. 1960. Leucine-induced hypoglycemia; a review and speculations. Am. J. Med. Sci. 240:792–801.

Earle, D. P., Jr., K. Smull, and J. Victor. 1942. Effects of excess dietary cysteic acid, *dl*-methionine, and taurine on the rat liver. J. Exp. Med. 76:317–323.

Elvehjem, C. A. 1956. Amino acid imbalance. Fed. Proc. 15:965–970.

Elvehjem, C. A., and W. H. Krehl. 1947. Imbalance and dietary interrelationships in nutrition. J. Am. Med. Assoc. 135:279–287.

Elvehjem, C. A., and W. H. Krehl. 1955. Dietary interrelationships and imbalance in nutrition. Borden's Rev. Nutr. Res. 16:69–84.

Fahey, J. L., D. Nathans, and D. Rairigh. 1957. Effect of L-arginine on elevated blood ammonia levels in man. Am. J. Med. 23:860–869.

Fajans, S. S., R. E. Knopf, J. C. Floyd, Jr., L. Power, and J. W. Conn. 1963. The experimental induction in man of sensitivity to leucine hypoglycemia. J. Clin. Invest. 42:216–229.

FN B (Food and Nutrition Board, National Research Council). 1959. Evaluation of protein nutrition. Publ. 711. National Academy of Sciences, Washington, D.C.

Goldstein, F. B. 1961. Biochemical studies on phenylketonuria. I. Experimental hyperphenylalaninemia in the rat. J. Biol. Chem. 236:2656–2661.

Gopalan, C. 1968. Leucine and pellagra. Nutr. Rev. 26:323–326.

Gopalan, C. 1970. Some recent studies in the Nutrition Research Laboratories, Hyderabad. Am. J. Clin. Nutr. 23:35–51.

Graham, C. E., S. W. Hier, H. K. Waitkoff, S. M. Saper, W. G. Bibler, and E. I. Pentz. 1950. Studies on natural and racemic amino acids with rats. J. Biol. Chem. 185:97–102.

Hall, C. D., E. A. Weiss, C. E. Morris, and A. J. Prange, Jr. Rapid deterioration in patients with Parkinsonism following tryptophan–pyridoxine administration. Neurology (Minneapolis) 22:231–237.

Hankes, L. V., J. E. Leklem, R. R. Brown, and R. C. P. M. Mekel. 1971. Tryptophan metabolism in patients with pellagra: Problem of vitamin B_6 enzyme activity and feedback control of tryptophan pyrolase enzyme. Am. J. Clin. Nutr. 24:730–739.

Harper, A. E. 1956. Amino acid imbalance, toxicities and antagonisms. Nutr. Rev. 14:225–227.

Harper, A. E. 1958a. Balance and imbalance of amino acids. Annu. N.Y. Acad. Sci. 69:1025–1041.

Harper, A. E. 1958b. Nutritional fatty livers in rats. Am. J. Clin. Nutr. 6:242–253.

Harper, A. E. 1965. Effect of variations in protein intake on enzymes of amino acid metabolism. Can. J. Biochem. 43:1589–1603.

Harper, A. E. 1966. Excesses of indispensable amino acids. Pages 221–228 *in* Toxicants occurring naturally in foods. Publ. No. 1354. National Academy of Sciences, Washington, D.C.

Harper, A. E. 1967. Effects of dietary protein content and amino acid pattern on food intake and preference. Pages 399–410 *in* Handbook of physiology. Vol. I. The American Physiology Society, Washington, D.C.

Harper, A. E. 1968. Amino acid balance and food intake regulation. Pages 181–217 *in* H. C. Meng and D. H. Law, eds. Parenteral nutrition: Proceedings of an international symposium, Vanderbilt University. C. C. Thomas, Springfield, Ill.

Harper, A. E. 1973. Amino acids of nutritional importance. Pages 130–152 *in* Toxicants occurring naturally in foods. 2nd ed. National Academy of Sciences, Washington, D.C.

Harper, A. E., D. A. Benton, and C. A. Elvehjem. 1955. L-leucine, an isoleucine antagonist in the rat. Arch. Biochem. Biophys. 57:1–12.

Harper, A. E., R. V. Becker, and W. P. Stucki. 1966. Some effects of excessive intakes of indispensable amino acids. Proc. Soc. Exp. Biol. Med. 121:695–699.

Harper, A. E., N. J. Benevenga, and R. M. Wohlheuter. 1970. Effects of ingestion of disproportionate amounts of amino acids. Physiol. Rev. 50:428–558.

Hegsted, D. M. 1963. Variation in requirements of nutrients—amino acids. Fed. Proc. 22:1424–1430.

Henkin, R. I., H. R. Keiser, and D. Bronzert. In press. Histidine dependent zinc loss, hypogensia, anorexia, and hyposmia. Gastroenterology. (A)

Hepburn, F. N., and W. B. Bradley. 1964. The glutamic acid and arginine requirement for high growth rate of rats fed amino acid diets. J. Nutr. 84:305–312.

Hepburn, F. N., and W. B. Bradley. 1968. Effect of glutamine on inhibition of rat growth by glycine and serine. J. Nutr. 94:504–510.

Hicks, C. S. 1926. Studies in tryptophane feeding. Aust. J. Exp. Biol. Med. Sci. 3:193–202.

Hier, S. W., C. E. Graham, and D. Klein. 1944. Inhibitory effect of certain amino acids on growth of young male rats. Proc. Soc. Exp. Biol. Med. 56:187–190.

Hodge, J. V., J. A. Oates, and A. Sjoerdsma. 1964. Reduction of the central effects of tryptophan by a decarboxylase inhibitor. Clin. Pharmacol. Ther. 5:149–155.

Holt, L. E., Jr. 1967. A Symposium on the Child, pp. 321–29. The Johns Hopkins Press, Baltimore.

Hsia, D. Y-Y. 1966. Inborn errors of metabolism. 2nd ed., Vol. 1. Yearbook Medical Publishers, Chicago, Ill. 396 pp.

Hueper, W. C., and G. J. Martin. 1943. Tyrosine poisoning in rats. Arch. Pathol. 35:685–694.

Hundley, J. M., H. R. Sandstead, A. G. Sampson, and G. H. Whedon. 1957. Lysine, threonine and other supplements to rice diets in man: Amino acid imbalance. Am. J. Clin. Nutr. 5:316–326.

Jones, J. D. 1964. Lysine–arginine antagonism in the chick. J. Nutr. 84:313–321.

Jones, J. D., R. Wolters, and P. C. Burnett. 1966. Lysine–arginine–electrolyte relationship in the rat. J. Nutr. 89:171–188.

Kennedy, J. L., W. Wertelecki, L. Gates, B. P. Sperry, and V. M. Cass. 1967. The early treatment of phenylketonuria. Am. J. Child. 113:16–21.

Kerr, G. R., and H. A. Waisman. 1967. Dietary induction of hyperphenylalaninemia in the rat. J. Nutr. 92:10–18.

Kerr, G. R., R. C. Wolf, and H. A. Waisman. 1966. A disorder of lipid metabolism associated with experimental hyperhistidinemia in *Macaca mulatta*. Pages 371–392 *in* Some recent developments in comparative medicine. Symp. Zool. Soc. (Lond.). Academic Press, New York.

Klavins, J. V., and P. V. Johansen. 1965. Pathology of amino acid excess. IV. Effects and interactions of excessive amounts of dietary methionine, homocystine and serine. Arch. Pathol. 79:600–614.

Knox, W. E., and O. Greengard. 1965. The regulation of some enzymes of nitrogen metabolism—an introduction to enzyme physiology. Adv. Enzyme Regul. 3: 247–313.

Krehl, W. A., L. J. Tepley, P. S. Sarma, and C. A. Elvehjem. 1945. Growth-retarding effect of corn in nicotinic acid-low rations and its counteraction by tryptophane. Science 101:489–490.

Lasser, R. P., M. R. Schoenfeld, and C. K. Friedberg. 1960. L-Lysine monohydrochloride: A clinical study of its action as a chloruretic acidifying adjuvant to mercurial diuretics. N. Engl. J. Med. 263:728–733.

Leung, P. M-B., and Q. R. Rogers. 1969. Food intake: Regulation by plasma amino acid pattern. Life Sci. 8(Part II):1–9.

Leung, P. M-B., Q. R. Rodgers, and A. E. Harper. 1968a. Effect of amino acid imbalance in rats fed *ad libitum*, interval-fed, or force-fed. J. Nutr. 95:474–482.

Leung, P. M-B., Q. R. Rogers, and A. E. Harper. 1968b. Effect of amino acid imbalance on dietary choice in the rat. J. Nutr. 95:483–492.

Lillie, R. D. 1932. Histopathologic changes produced in rats by the addition to the diet of various amino acids. U.S. Public Health Rep. 47:83–93.

Linkswiler, H., H. M. Fox, and P. C. Fry. 1960. Availability to man of amino acids from foods. IV. Isoleucine from corn. J. Nutr. 72:397–403.

Lucas, R. R., and J. P. Newhouse. 1957. The toxic effects of sodium glutamate on the inner layers of the retina. AMA Arch. Ophthalmol. 58:193–201.

Martin, G. J., and W. C. Hueper. 1943. Biochemical lesions produced by diets high in tyrosine. Arch. Biochem. 1:435–438.

McClellan, W. S., and E. F. Dubois. 1930. Clinical calorimetry. XIV. Prolonged meat diets with a study of kidney function and ketosis. J. Biol. Chem. 87:651–668.

Merimee, T. J., D. A. Lillicrap, and D. Rabinowitz. 1965. Effect of arginine on serum-levels of human growth hormone. Lancet 2:668–670.

Milne, M. D. 1968. Pharmacology of amino acids. Clin. Pharmacol. Ther. 9:484–514.

Muramatsu, K., and K. Ashida. 1962. Effect of dietary protein level on growth and liver enzyme activities in rats. J. Nutr. 76:143–150.

Naber, E., W. W. Cravens, C. A. Baumann, and H. R. Bird. 1956. The relation of dietary supplements and tissue metabolites to glycine toxicity in the chick. J. Nutr. 60:75–85.

Najarian, J. S., and H. A. Harper. 1956. A clinical study of the effect of arginine on blood ammonia. Am. J. Med. 21:832–842.

Nakagawa, I., T. Takahashi, T. Suzuki, and K. Kobayashi. 1964. Amino acid requirements of children: Nitrogen balance at the minimal level of essential amino acids. J. Nutr. 83:115–118.

Nesheim, M. C. 1968a. Genetic variation in arginine and lysine utilization. Fed. Proc. 27:1210–1214.

Nesheim, M. C. 1968b. Kidney arginase activity and lysine tolerance in strains of chickens selected for a high or low requirement of arginine. J. Nutr. 95:79–87.

Nyhan, W. L. 1967. Amino acid metabolism and genetic variation. McGraw-Hill, New York. 495 pp.

Oates, J. A., and A. Sjoerdsma. 1960. Neurologic effects of tryptophan in patients receiving a monamine oxidase inhibitor. Neurology 10:1076–1078.

O'Dell, B. L., and W. O. Regan. 1962. Effect of lysine and glycine upon arginine requirement of guinea pigs. Proc. Soc. Exp. Biol. Med. 112:336–337.

Ogden, D. A., L. Scherr, N. Spritz, A. L. Rubin, and E. H. Luckey. 1961. The management of resistant fluid-retention states with intravenous L-arginine monohydrochloride in combination with mercurial diuretics. Am. Heart J. 61:16–20.

Olney, J. W. 1969. Brain lesions, obesity and other disturbances in mice treated with monosodium glutamate. Science 164:719–721.

Olney, J. W., and L. G. Sharpe. 1969. Brain lesions in an infant rhesus monkey treated with monosodium glutamate. Science 166:386–388.

Orr, M. L., and B. K. Watt. 1957. Amino acid content of foods. USDA, Home Econ Rep. No. 4. 82 pp.

Oser, B. L. 1945. A method for integrating essential amino acid content in nutritional evaluation of protein. J. Am. Diet. Assoc. 27:396–402.

Peng, Y., and A. E. Harper. 1970. Amino acid balance and food intake: Effect of different dietary amino acid patterns on plasma amino acid pattern of rats. J. Nutr. 100:429–437.

Peng, Y., J. K. Tews, and A. E. Harper. 1972. Amino acid imbalance, protein intake, and changes in rat brain and plasma amino acids. Am. J. Physiol. 222:314–321.

Persaud, T. V. 1969. The foetal toxicity of leucine in the rat. West Indian Med. J. 18:34–39.

Phear, E. A., B. Ruebner, S. Sherlock, and W. H. Summerskill. 1956. Methionine toxicity in liver disease and its prevention by chlortetracycline. Clin. Sci. 15:93–117.

Polidora, V. J. 1967. Behavioral effects of "phenylketonuria" in rats. Proc. Nat. Acad. Sci. 57:102–106.

Pollin, W., P. V. Cardon, Jr. and S. S. Kety. 1961. Effects of amino acid feedings in schizophrenic patients treated with iproniazid. Science 133:104–105.

Rama Rao, P. B., V. C. Metta, H. W. Norton, and B. C. Johnson. 1960. The amino acid composition and nutritive value of proteins. III. The total protein and the nonessential amino nitrogen content. J. Nutr. 71:361–365.

Reynolds, T. B., A. G. Redeker and P. Davis. 1958. A controlled study of the effects of L-arginine on hepatic encephalopathy. Am. J. Med. 25:359–367.

Rose, W. C. 1957. The amino acid requirements of adult man. Nutr. Abstr. Rev. 27:631–647.

Rosenberg, H. R., R. Culik, and R. E. Eckert. 1959. Lysine and threonine supplementation of rice. J. Nutr. 69:217–228.

Ross, M. H. 1959. Protein, calories and life expectancy. Fed. Proc. 18:1190–1215.

Ross, M. H. 1961. Length of life and nutrition in the rat. J. Nutr. 75:197–210.

Russell, W. C., M. W. Taylor, and J. M. Hogan. 1952. Effect of excess essential amino acids on growth of the white rat. Arch. Biochem. Biophys. 39:249–253.

Salmon, W. D. 1958. The significance of amino acid in nutrition. Am. J. Clin. Nutr. 6:487–494.

Sauberlich, H. E. 1961. Studies on the toxicity and antagonism of amino acids for weanling rats. J. Nutr. 75:61–72.

Schimke, R. T. 1963. Studies on factors affecting the levels of urea cycle enzymes in rat liver. J. Biol. Chem. 238:1012–1018.

Schweizer, W. 1947. Studies on the effect of L-tyrosine on the white rat. J. Physiol. 106:167–176.

Scrimshaw, N. S., R. Bressani, M. Béhar, and F. Viteri. 1958. Supplementation of cereal proteins with amino acids. I. Effect of amino acid supplementation of corn-masa at high levels of protein intake on the nitrogen retention of young children. II. Effect of amino acid supplementation of corn-masa at intermediate levels of protein intake on the nitrogen retention of young children. J. Nutr. 66:485–514.

Sidransky, H., and E. Farber. 1958. Chemical pathology of acute amino acid deficiencies. II. Biochemical changes in rats fed threonine- or methionine-devoid diets. Arch. Pathol. 66:135–149.

Snyderman, S. E., E. L. Holt, Jr., P. M. Norton, and E. Roitman. 1968. Effect of high and low intakes of individual amino acids on the plasma aminograms. Pages 19–31 in J. H. Leathem, ed. Protein nutrition and free amino acid patterns. Rutgers University Press, New Brunswick, N.J.

Solomons, C. C., E. K. Cotton, R. Dubois, and M. Pinney. 1971. The use of buffered L-arginine in the treatment of cystic fibrosis. Pediatrics 47:384–390.

Spolter, P. D., and A. E. Harper. 1961. Leucine–isoleucine antagonism in the rat. Am. J. Physiol. 200:513–518.

Stanbury, J. B., J. B. Wyngaarden, F. Hanes, and D. S. Fredrickson. 1972. The metabolic basis of inherited disease. 3rd ed. McGraw-Hill, New York. 2240 pp.

Sullivan, M. X., W. C. Hess, and W. H. Sebrell. 1932. Studies on the biochemistry of sulfur. XII. Preliminary studies on amino-acid toxicity and amino-acid balance. U.S. Public Health Rep. 47:75–83.

Van Pilsum, J. F., and C. P. Berg. 1950. The comparative availabilities of mixtures of the L and DL modifications of the essential amino acids for growth in the rat. J. Biol. Chem. 183:279–290.

Waisman, H. A., and H. F. Harlow. 1965. Experimental phenylketonuria in infant monkeys. Science 147:685–695.

Watt, B. K., and A. L. Merrill. 1963. Composition of food. USDA Agric. Handb. No. 8. 190 pp.

Willcock, E. G., and F. G. Hopkins. 1906. The importance of individual amino acids in metabolism; observations on the effect of adding tryptophane to a dietary in which zein is the sole nitrogenous constituent. J. Physiol. 35:88–102.

Wretlind, K. A. J. 1952. The availability for growth and the toxicity of L- and D-phenylalanine. Acta Physiol. Scand. 25:276–285.

Yoshida, A., K. Ashida, and A. E. Harper. 1961. Prevention of fatty liver due to threonine deficiency by moderate caloric restriction. Nature 189:917–918.

Yoshida, O., R. R. Brown, and G. T. Bryan. 1971. A possible role of urinary metabolites of tryptophan in the heterotopic recurrence of bladder cancer in man. Am. J. Clin. Nutr. 24:848–851.

Yuwiler, A., and R. T. Louttit. 1961. Effects of phenylalanine diet on brain serotonin in the rat. Science 134:831–832.

Zimmerman, R. A., and H. M. Scott. 1965. Interrelationship of plasma amino acid levels and weight gain in the chick as influenced by suboptimal and superoptimal dietary concentrations of single amino acids. J. Nutr. 87:13–18.

E. F. PHIPARD

Protein and Amino Acids in Diets

PROTEIN INTAKES—UNITED STATES

The amount of protein provided by the U.S. food supply has usually averaged between 90 and 100 g per capita/day since 1909–13, the earliest recorded period (Table 1). These estimates were obtained by applying appropriate food composition values to the per capita quantities of individual foods available for consumption at the retail stage of distribution. Although these figures are not measures of intake, they do show trends in overall patterns of consumption. Contrary to common belief, protein consumption has not increased over the years, but the relative amounts of protein from animal and plant sources have changed. Today more than two-thirds comes from animal sources— meat, poultry, fish, eggs, milk, and cheese. Sixty or more years ago, the proportion from these sources was about half, reflecting greater consumption of grain products at that time.

Average amounts of eight essential amino acids and of total sulfur-containing acids provided by the per capita food supply for 1970 and the percentage of each contributed by major food groups are shown in Table 2. These quantities are generous in terms of estimated human requirements; the sulfur-containing amino acids, which are relatively the least abundant, are present at roughly three to four times the estimated

TABLE 1 Sources of Protein in U.S. Food Supply (Selected Years)

Foods	Percentage of Protein from Specified Sources							
	1909–1913	1925–1929	1935–1939	1947–1949	1957–1959	1965	1970
Dairy products (including butter)	16.5	19.1	21.2	23.7	24.6	23.7	22.1
Eggs	5.2	6.1	5.8	7.1	6.8	5.9	5.8
Meat, poultry, fish	29.9	29.4	29.3	32.7	35.7	38.7	41.5
Total animal sources[a]	51.7	54.7	56.3	63.6	67.0	68.3	69.4
Grain products	35.8	31.8	28.7	22.9	19.9	19.0	18.1
Legumes, nuts	4.5	4.8	5.7	5.0	5.2	5.1	5.0
Other vegetable sources	8.0	8.8	9.4	8.4	7.8	7.6	7.5
Total vegetable sources[a]	48.3	45.3	43.7	36.4	33.0	31.7	30.6
Protein/person/day (g)	102	95	90	95	95	97	100

[a]Components may not add to totals due to rounding.
SOURCES: USDA, 1968, 1970, 1971. Calculations were made by the Consumer and Food Economics Research Division, ARS, USDA.

TABLE 2 Essential Amino Acids in U.S. Food Supply (1970) and Percentage Contribution of Major Food Groups to Total[a]

Amino Acid	Amount per Capita per Day (mg)	Percent of Total from Specified Food Groups					
		Dairy Products	Eggs	Meat, Poultry, Fish	Grain Products	Legumes, Nuts	Other Vegetable Sources
Isoleucine	5,300	26.6	7.2	40.4	15.7	5.0	5.2
Leucine	8,200	26.6	6.2	40.0	17.8	4.8	4.6
Lysine	6,700	25.6	5.5	52.5	6.8	4.3	5.3
Methionine	2,100	25.4	8.4	48.3	11.6	2.8	3.5
Total sulfur-containing	3,500	21.2	9.1	44.6	17.3	3.9	3.9
Phenylalanine	4,700	23.4	7.1	36.2	20.6	5.9	6.8
Threonine	4,100	24.1	7.0	44.3	13.7	4.6	6.4
Tryptophan	1,200	24.7	7.8	39.0	17.4	4.9	6.3
Valine	5,700	27.0	7.6	39.1	14.9	5.2	6.3

[a]Data computed by Consumer and Food Economics Research Division, ARS, USDA, based on estimates of per capita consumption (retail basis) supplied by ERS, USDA.

human requirements. Major contributors are meat, poultry, and fish; dairy products; and grain products, in that order.

Studies of household food consumption and of food intakes of individuals indicate how the food supply is distributed among the population. Most surveys show a wide range in the distribution of all nutrients, including protein, and in energy, to which protein supplies are related. Some of the variability is related to food energy requirements, which influence total food intake, and to food selection; some is related to income and the amount of money available for food.

Households

A nationwide survey of household diets in spring 1965 (USDA, 1969) indicated that the foods estimated to have been used in the household during a week provided an average of 106 g of protein per person per day for the nation as a whole. Averages for different income classes are as follows:

Income	Protein per person per day (g)
Under $3,000	98
$3,000–$4,999	102
$5,000–$6,999	107
$7,000–$9,999	110
$10,000 and over	113

Among individual households, a wide range in dietary protein supply was found at all income levels (Table 3). On a nationwide basis, 95 percent had diets supplying at least the amount of protein specified in the NAS–NRC allowances for individuals of different sex and age (FNB, 1964). Of households with incomes under $3,000, 88 percent had diets providing recommended amounts, 2 percent had diets providing less than two-thirds of the allowances, and 10 percent had diets providing between two-thirds and 100 percent of allowances.

Included in Table 3 are data from several small studies of families with limited incomes. One of these is a study of one week's food consumption of one- and two-person households in Rochester, New York, in which one person was receiving old age, survivors, and disability insurance. Included also are data from studies, made in five localities in the early 1960's, of families with incomes low enough in relation to family size to qualify them for USDA food assistance programs but who were not participating in them. Although all were eligible, they were classified in three relative income categories for purposes of analysis. In 1967, similar studies were made of diets of low-income families in two counties of the Mississippi delta that were participating, or eligible to participate, in USDA food programs. Protein supplies were comparable to those in the earlier studies.

As might be expected, a larger proportion of those having diets with low-protein levels were found among the lower income groups than in the population as a whole. What may be surprising is the comparatively large proportion with high protein intakes.

TABLE 3 Protein: Percent of Household Diets in a Week Providing Specified Proportions of the Recommended Dietary Allowances[a]

Study and Income Class	Households (No.)	Percent Less Than 67% RDA	Percent 67–99% RDA	Percent 100% RDA or More
Nationwide, 1965[b]				
All urbanizations[c]	6,174	1	4	95
Under $3,000	1,294	2	10	88
$3,000–$4,999	1,157	1	4	95
$5,000–$6,999	1,483	h	4	96
$7,000–$9,999	1,219	h	2	98
$10,000 and over	685	0	2	98
OASDI study, Rochester[d]	283	2	17	83
USDA food program studies				
5 localities, nonparticipants[e]				
Low income I (lowest)[f]	176	7	15	78
Low income II[f]	195	6	20	74
Low income III[f]	248	2	10	88
Mississippi survey, 1967[g]				
Food donation program				
Participants	145	2	19	79
Nonparticipants	44	7	18	75
Food stamp program				
Participants	117	4	11	85
Nonparticipants	178	7	18	75

[a]FNB, 1964, for 1965 and 1967 surveys; FNB, 1958, for other surveys.
[b]USDA, 1969.
[c]Includes households not classified by income.
[d]One- and two-person households with one person receiving old age, survivors and disability insurance. Unpublished data, CFE Division, ARS, USDA.
[e]Families of two or more persons that qualified for food assistance but were not participating in programs. Unpublished data, CFE Division, ARS, USDA.
[f]Based on extent to which family income met the income standard set by the state for participation in food assistance programs.
[g]Unpublished data, CFE Division, ARS, USDA.
[h]0.5 or less.

Individuals

Information on the food intake of individuals is essential for understanding how food supplies are actually utilized by consumers. For the
first time data on food consumption and nutrient intake for a random
sample of individuals in the United States are available. They were
obtained as a part of the USDA's nationwide survey of household food
consumption in 1965.

Average energy and protein intakes of 14,500 individuals, grouped
by sex and age, are shown in Table 4. These figures, based on recalls of
food eaten in a single day, are believed to be indicative of usual consumption by specified sex and age groups. Average protein intakes are
generous, exceeding the NAS–NRC recommended dietary allowances
(FNB, 1968) by a considerable margin. Average energy intakes of males

TABLE 4 Average Energy and Protein Content of 1-Day Diets of Individuals,
by Sex and Age

Sex–Age Group	No. of Persons	Food Energy[a] (kcal)	Protein[b] (g)
Male and female:			
Under 1 yr[c]	408	960	39
1–2 yr	810	1,405	56
3–5 yr	1,405	1,705	65
6–8 yr	1,412	2,015	76
Male:			
9–11 yr	665	2,355	88
12–14 yr	627	2,660	100
15–17 yr	562	2,990	114
18–19 yr	251	3,050	118
20–34 yr	1,406	2,915	119
35–54 yr	2,050	2,630	106
55–64 yr	742	2,420	98
65–74 yr	460	2,060	82
75 yr and over	219	1,870	73
Female:			
9–11 yr	599	2,010	75
12–14 yr	626	2,145	81
15–17 yr	538	2,000	78
18–19 yr	232	1,920	76
20–34 yr	1,846	1,805	72
35–54 yr	2,492	1,650	68
55–64 yr	916	1,620	67
65–74 yr	624	1,475	60
75 yr and over	340	1,460	59

[a]Calories rounded to nearest 5.
[b]Protein rounded to nearest gram.
[c]Does not include nursing infants. In the total sample, 15 such infants were reported.
SOURCE: USDA, 1972.

are close to those recommended; those of females are considerably lower. Group averages such as these are useful for many purposes, but they reveal little about the prevalence of nutritionally poor diets. They need to be supplemented by more detailed information on the proportions of the population whose diets provide different nutrient levels.

Information on the food intake of more than 500 preschool children in Mississippi was obtained in the winter of 1967–68 as a part of a study of nutritional status (Owen *et al.*, 1969). This study was designed to assess the extent and severity of malnutrition in a state known to have a high proportion of low-income families. Nutrient intake data were classified by income and by certain other factors for analysis. Children in the lowest annual income class (<$500 per capita) had lower intakes of most nutrients than did those from families with higher incomes. Average energy intakes ranged from 80 kcal/kg of body wt in the lowest of four income groups to 101 kcal/kg in the highest group. Protein intakes averaged 3 g/kg in the lowest income classes and 3.8 g/kg in the highest.

Among individual children in two income groupings, the proportions with intakes of energy and protein considered "low" and "deficient" were as follows:

	Income groups	
	<$500	>$500
	(Percent)	(Percent)
Low intakes		
kcal < 75/kg	44	24
Protein <1.5 g/kg	6	3
Deficient intakes		
kcal < 60/kg	29	9
Protein <1.2 g/kg	4	1

The data show that a small proportion of children had diets low or deficient in protein. In fact, 95 percent of the values were between 2.6 and 3.9 g/kg. With energy intake frequently low, physiologic protein deficiencies might have been anticipated; however, clinical examination revealed only one child who was judged to have some evidence of protein–calorie malnutrition. Inadequate intakes of several other nutrients were found, especially in the lower income groups.

Additional information on diets of individuals is available from the Ten-State Nutrition Survey (USDHEW, 1972). Within each state the population studied was selected from areas showing the greatest amount of poverty according to the 1960 Census.

In general, protein was one of the nutrients that appeared to be well supplied. Mean protein intakes of most age, sex, and ethnic groups, as calculated from 1-day food recalls, met or exceeded dietary standards. However, when data for individuals were arrayed by cumulative percentage distributions, a large number of persons in most analysis groups were found to have less than recommended intakes of protein. For example, about one-fifth of the females aged 12–14 and 15–16 in the five lower-income states had less than 40 g of protein in their food for one day. In these states also, 38 percent of the pregnant women had intakes of less than 50 g. Many persons over 60, especially women, had low protein intakes, frequently associated with low total food consumption. It is, in fact, often difficult to meet the body's need for nutrients while at the same time restricting calories in line with today's reduced level of activity.

A number of studies suggest that family members do not necessarily share in household supplies of protein in accordance with their nutritional needs. In families where money for food is limited, the father and older boys may well be given a disproportionate share of high-protein foods. Also, low protein intakes of some family members may be the result of food preferences and whims, or lack of information about foods and food needs.

An increasingly important cause for U.S. diets being low in protein is the reduced food intake associated with sedentary living and with advancing age. This was revealed in studies of diet and nutritional status of women in the north central region of the United States. Swanson *et al.* (1959) reported that "For every increase of 10 years in age, the nutritive value of diets of Iowa women decreased on the average by about 85 calories, 4 g protein, 0.03 g calcium, 1.4 mg ascorbic acid, and 194 I.U. vitamin A."

Diets of South Dakota women showed a trend similar to that in Iowa (Burrill and Alsup, 1955). They provided daily, on the average, amounts of energy and protein for various age classes as follows:

Age class	No.	Energy (kcal)	Protein (g)
30–39	91	1,840	64
40–49	105	1,770	59
50–59	73	1,608	55
60–69	45	1,630	52
70 and over	25	1,353	42

Evaluation of these nutrient intakes is provided by balance studies done on some of the women participating in the north central studies

(Ohlson *et al.*, 1952). At each age–decade beyond the thirties, mean nitrogen retentions were negative (Figure 1), even though intakes were of the order of 8 to 10 g (50 to 60 g protein). Figure 1 also shows a decreasing caloric intake with increasing age. When these balance data were classified by caloric intake level, mean nitrogen retentions were negative for all age groups whenever energy intakes were under 1,500 kcal. At least 1,800 kcal/day were required to assure nitrogen equilibrium in this group of women.

As a number of studies have shown, many persons, particularly women, consume less protein than is recommended. However, the NAS–NRC allowances for protein are believed to provide generously for the needs of most individuals, and consequently intakes below recommended amounts are not necessarily inadequate for a particular individual. Whether or not protein requirements are met depends not only on total nitrogen intake in relation to body size but upon the amounts of the various amino acids in the food—meal-by-meal, total energy value, and perhaps other factors.

The essential amino acid content of diets of 10 individuals is shown in Table 5. These 1-day diets of homemakers were chosen because studies have shown homemakers to be one of the groups more likely to have poor diets. As it turned out, the diets selected represent a comparatively low level of protein intake, 40 g, and were poor in many other respects by present-day standards. Apparently, however, they

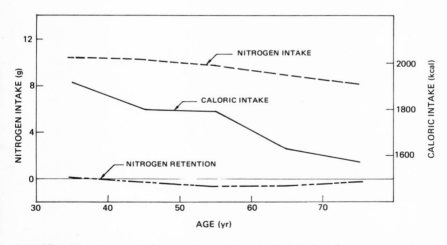

FIGURE 1 Mean nitrogen intakes, retentions, and mean caloric intakes of women, successive decades. From Ohlson *et al.*, 1952.

TABLE 5 Homemakers' Diets: Caloric, Niacin, and Amino Acid Content of Selected 1-Day Diets Providing 40 g of Protein

Diet Number	Subject Weight (kg)	Number kcal	Niacin (mg)	Isoleucine (g)	Leucine (g)	Lysine (g)	Methionine (g)	Cystine (g)	Total Sulfur A.A. (g)	Phenylalanine[a] (g)	Threonine (g)	Tryptophan (g)	Valine (g)
Birmingham													
1	49	1,280	8.5	2.2	3.2	2.3	0.9	0.6	1.5	2.0	1.6	0.5	2.3
2	64	1,120	11.2	2.1	3.3	2.9	0.9	0.6	1.5	1.8	1.7	0.4	2.2
3	65	2,325	11.1	2.0	3.3	2.0	0.8	0.8	1.5	2.2	1.6	0.5	2.4
4	55	2,325	7.1	1.8	3.5	2.5	0.6	0.6	1.2	2.0	1.5	0.3	2.2
5	55	2,500	7.2	2.2	3.7	2.6	0.6	0.5	1.1	2.2	1.7	0.4	2.5
Minneapolis													
6	77	1,360	9.1	2.4	3.5	3.1	1.0	0.6	1.6	2.1	1.9	0.6	2.6
7	50	1,540	9.0	2.2	3.4	2.7	0.9	0.7	1.6	2.0	1.7	0.5	2.4
8	70	1,720	11.4	2.2	3.4	2.8	0.9	0.7	1.6	2.0	1.7	0.5	2.3
9	61	1,590	6.8	2.1	3.0	2.7	0.8	0.6	1.4	1.7	1.6	0.4	2.2
10	54	1,300	6.8	2.0	3.1	2.2	0.8	0.6	1.4	1.8	1.5	0.5	2.2

[a]Total aromatic amino acids were not computed.
SOURCE: Unpublished data from 1948 Study of Urban Family Food Consumption. Institute of Home Economics, U.S. Department of Agriculture. Reprinted from FNB, 1959, Table 10.

provided at least the estimated minimum amount of each essential amino acid required by normal adult women, but whether or not these women were in positive nitrogen balance was not determined. That they were seems unlikely in view of the low energy intakes of some of them.

PROTEIN AND ENERGY CONTENT OF DIETS IN COUNTRIES OTHER THAN THE UNITED STATES

Diets in few countries of the world are as well supplied with energy, protein, and other nutrients as those in the United States. Information on the food supply of other countries is compiled by the Food and Agriculture Organization (FAO) as part of its continuing study of the state of food and agriculture. It also originates in the USDA's statistical studies of food supplies and food needs of the countries of the world, such as those summarized in *The World Food Budget, 1970* (USDA, 1964). In order to identify the food deficit areas of the world for this study, the USDA adopted nutrition reference standards as a basis for arbitrarily classifying countries as "diet-adequate" or "diet-deficit." The reference standards for energy were in line with those developed by FAO for short-term targets in the Third World Food Survey (FAO, 1963). They ranged from 2,300 to 2,700 kcal/person/day for different areas. A standard for protein was included also, since the amount and type of protein consumed is considered to be an indicator of the nutritional quality of the whole diet. Not only is protein a basic dietary essential, but protein-rich foods, especially those of animal origin, are good sources of certain other essential nutrients. The reference standard adopted for protein, for all countries, was 60 g/person/day, of which at least 10 g were to be from animal sources and another 10 g from animal sources and/or pulses. For fat, the reference standard was arbitrarily set as the amount to provide 15 percent of the reference standard for energy.

Of 92 countries for which so-called balance sheets were prepared for the World Food Budget, 37 had food supplies in 1959–61 providing fewer calories per capita than were specified in the nutrition reference standards. Protein supplies were below the reference standard in 33 countries. Quality was probably poor also, as judged by how small a proportion was derived from animal sources or from pulses. Many of the countries with low protein supplies also had low energy supplies. The inhabitants of countries with food supplies that are by definition deficient in energy and/or protein account for a large proportion of the world's population. This is not to say that other countries do not encounter problems in connection with nutrient supplies. In Brazil, for example, energy and protein supplies for the country met the reference standards; but in large areas either the quality or quantity of the food, or both, were inadequate, and malnutrition was widespread. In most countries, distribution of the food supply is not equitable. Thus that segment of the population far from the source of supplies, or which has little money, may fare badly.

When the amount of protein available per capita is around 40–50 g, as it is in several countries in the Far East, parts of Africa, and Central America (Table 6), the proportion of individuals with low-protein diets must necessarily be large. Furthermore, when low total protein is associated with a low energy supply the situation is even worse than it appears, since some of the protein will probably be used to meet energy needs.

The four countries listed last in Table 6 are included as examples of places where food supplies are extremely low in protein but have energy levels that met the reference standard. Important sources of energy in these countries are starchy roots and sugars, which furnish little or no protein.

High carbohydrate foods—grains, starchy roots, sugar—account for 70–80 percent of the energy in many food-deficit countries. Because they contain varying amounts of protein, the kinds and amounts of

TABLE 6   Energy and Protein Levels Per Capita in Food Supplies of Selected Countries, 1959–1961

Country	Energy (kcal)	Protein (g) Total	Animal	Pulse
British Honduras	2,020	53	14	4
Dominican Republic	2,020	41	12	7
Guatemala	1,970	50	9	4
Haiti	1,780	46	7	15
Peru	2,060	51	13	5
Burma	2,170	46	9	4
Ceylon	2,120	47	12	6
Indonesia	2,160	43	5	8
Philippines	2,000	45	11	4
Thailand	2,120	45	9	4
Congo	2,460	44	4	15
Guinea	2,400	46	3	5
Liberia	2,430	39	5	4
Sierra Leone	2,410	40	3	3

SOURCE: USDA, 1964.

these foods that serve as the dietary staple can make a substantial difference in the total protein and amino acid content of the diet. To illustrate, 1,500 kcal from the following foods provide very different amounts of protein and of certain key amino acids:

Food	Protein (g)	Tryptophan (g)	Lysine (g)	Total sulfur-containing (g)
Cassava	7	0.094	0.295	0.125
Rice	28	0.343	1.254	1.003
Millet A	29	0.379	0.901	2.039
Cornmeal	39	0.235	1.111	1.216
White flour	45	0.533	0.987	1.437
Millet B	45	0.591	1.373	1.827
Oatmeal	55	0.709	2.018	2.006
Whole wheat	62	0.777	1.725	2.341

A meaningful appraisal of a nation's food supply requires more information than is provided by balance sheet averages. The real test of the adequacy of a diet is the nutritional well-being of the person who consumes it. In many of the developing countries, various groups have conducted at least limited studies of the food consumption, health, and nutritional status of selected portions of the population. Many of these studies have been made under country auspices, some with assistance from international agencies; and others were made in cooperation with survey teams of the Interdepartment Committee on Nutrition for National Defense.

One cannot generalize about results of these studies, even for a given country, except to say that one or more dietary deficiencies are usually found, especially among children. Among adults signs of protein malnutrition are less common than those of deficiencies of certain other nutrients.

Although dietary levels appear low in terms of commonly accepted recommended allowances for protein, they may not be low in relation to body size in some populations. For example, the food supply of Burma in 1959–61 provided 46 g of protein and 2,170 kcal/person/day (Table 6). Average heights and weights of the civilian adults who were measured during a nutrition survey in Burma in 1961 (U.S. Interdepartmental Committee on Nutrition for National Defense, 1963) were:

	Height (cm)	Weight (kg)
Males 15+ yr	161.5	49.3
Females 15+ yr (non-pregnant, non-lactating)	150.7	42.8

These measurements suggest that the protein supply (46 g/person) was sufficient to provide an intake of 1 g/kg of body wt, a level that has been considered adequate, except for young children, to compensate for a low Biological Value of the dietary protein, provided energy needs are met. With the low energy intake level in Burma, however, the average protein supply, 46 g per capita, must be borderline at best, even for people of small stature and even if the food is equitably distributed. Small body size, as Waterlow (1968) has suggested, represents an effective physiological adaptation to inadequate food.

Although signs of protein deficiency were not seen in the Burma nutrition survey, the low consumption of animal products undoubtedly contributed to the deficiencies of thiamin, riboflavin, and iron frequently seen in the clinical and biochemical studies. This situation is common in other countries as well. The importance of animal products in reinforcing the diet, not only in protein but also in other nutrients often deficient, should not be overlooked in plans and programs for improving the nutrition of underfed populations.

PROTEIN IN DIETS OF YOUNG CHILDREN

The most critical nutrition problem in much of the world is "protein-energy" malnutrition of young children. This problem is not confined to the diet-deficit areas indicated in the World Food Budget, but occurs in most of the countries of Asia, Africa, and South and Central America. Although quantitative information on its prevalence is lacking, protein–energy deficiency has been estimated (Patwardhan, 1964) to be acute in 2–10 percent of infants and young children and to be present in milder forms in a much larger proportion, especially in the populous developing countries. The basic cause is the inadequacy of the post-weaning diet given to the child during the time it is exposed recurrently to infection. This diet is likely to be low in protein, to provide insufficient energy, and often to be marginal in other nutrients as well. Poverty is an underlying factor. However, distribution of the food among family members is determined largely by the cultural pattern, and the young child frequently fares worst.

Children in Nigeria, for example, may not share in the family food supply in accordance with their nutritional needs. This was shown in a study of the diets of urban dwellers in Lagos (McFie, 1967). Mean nutrient intakes of energy and protein by different age or sex groups were compared to recommended allowances, (FAO, 1957a,b), as shown below:

Sex or Age Group	Energy (kcal)	Percent of Allowances	Protein (g)	Percent of Allowances
Adult males	2,010	77	71	128
Adult females	1,610	78	54	108
Pregnant and lactating women	2,062	66	63	65
Children				
10–12 yr	1,141	55	31	63
7–9 yr	1,207	61	33	83
4–6 yr	1,134	68	30	94

The data show that children's intakes, relative to allowances, were lower than those of adult males and females. In this instance the younger children fared better than those 10–12 yr old. The figures show also that energy intakes of all groups except pregnant and lactating women were lower relative to allowances than were protein intakes.

The protein content of diets reflects to a great extent the kind and composition of the dietary staple. The cereal-based diets of northern groups in Nigeria, for example, provide more protein in relation to energy than do those in the south, where starchy roots are staples.

Extensive studies (Rao *et al.*, 1969) of the diets of preschool children in India provide information on relative deficiencies of protein and energy in this age group. Mean protein intakes in six regions ranged from 19–28 g, or 1.7–2.8 g/kg body weight. These amounts are above the 1.40 g/kg recommended by the Indian Council of Medical Research in 1968. Mean energy intakes were 600–950 kcal for the same groups, which is well below the recommended allowance of 1,275 kcal for children of 1–5 yr.

Further analysis of the data from one region showed that 35 percent of the children had less than recommended intakes of both protein (body weight basis) and energy; 57 percent had enough protein, but were deficient in energy intake. No diets were found in this group that were adequate in energy value but deficient in protein. The authors conclude that nearly 90 percent of these children probably were victims of some degree of protein–energy malnutrition, primarily the result of an energy deficiency. They add that if the children were to "consume the types of diets they are eating now in adequate amounts to satisfy their energy needs, their protein needs would be satisfied." It would be wasteful, they suggest, to provide protein concentrates as supplements to the children's usual diets without at the same time assuring adequate energy intake.

A study of the food intake of preschool Guatemalan Indian children (Flores *et al.*, 1966) provides an example of nutrition problems in

Latin American countries. Average energy and protein intakes of four age groups in each of three villages are given below. The N A S–N R C Recommended Dietary Allowances (R D A 's) (F N B, 1968) for these age groups are provided for comparison.

	1–2 yr	2–3 yr	3–4 yr	4–5 yr
Village 1				
Kcalories	498	883	1,065	1,215
Protein (g/day)	15.0	25.8	32.1	34.9
Village 2				
Kcalories	417	869	978	1,110
Protein (g/day)	11.9	22.8	27.1	31.4
Village 3				
Kcalories	525	723	847	1,023
Protein (g/day)	12.8	18.9	21.9	26.7
U.S.–RDA				
Kcalories	1,100	1,250	1,400	1,600
Protein (g/day)	25	25	30	30

Compared to R D A 's in the United States, energy intakes were low in all groups. Protein intakes were low in many groups, especially the 1–2-yr-olds, and in all age groups in Village 3. Because of energy deficits, protein shortages must have been even more serious than these figures indicate.

The foregoing information is cited to illustrate how low the protein and energy intakes of young children in three developing countries are. Many children are even less well fed than the ones shown here and protein–energy malnutrition of young children is but one aspect of the larger problem of food shortages.

REFERENCES

Burrill, L. M., and B. Alsup. 1955. Food habits of South Dakota women. S. Dak. Agric. Exp. Stn. Bull. 451. 24 pp.

F A O (Food and Agriculture Organization). 1957a. Calorie requirements. F A O Nutr. Stud. No. 15. F A O, Rome. 67 pp.

F A O (Food and Agriculture Organization). 1957b. Protein requirements. F A O Nutr. Stud. No. 16. F A O, Rome. 52 pp.

F A O (Food and Agriculture Organization). 1963. Third world food survey. Freedom from Hunger Campaign Basic Study No. 11. F A O, Rome. 102 pp.

Flores, M., Z. Flores, and M. Y. Lara. 1966. Food intake of Guatemalan Indian children, ages 1 to 5. J. Am. Diet. Assoc. 48:480–487.

F N B (Food and Nutrition Board, National Research Council). 1958. Recommended dietary allowances. Revised ed. Publ. 589. National Academy of Sciences, Washington, D.C. 36 pp.

F N B (Food and Nutrition Board, National Research Council). 1959. Evaluation of protein nutrition. Rep. 711. National Academy of Sciences, Washington, D.C. 61 pp.

FNB (Food and Nutrition Board, National Research Council). 1964. Recommended dietary allowances. 6th ed. Publ. 1146. National Academy of Sciences, Washington, D.C. 59 pp.

FNB (Food and Nutrition Board, National Research Council). 1968. Recommended dietary allowances. 7th ed. Publ. 1694. National Academy of Sciences, Washington, D.C. 101 pp.

McFie, J. 1967. Nutrient intakes of urban dwellers in Lagos, Nigeria. Br. J. Nutr. 21:257–268.

Ohlson, M. A., W. D. Brewer, L. Jackson, P. P. Swanson, P. H. Roberts, M. Mangel, R. M. Leverton, M. Chaloupka, M. R. Gram, M. S. Reynolds, and R. Lutz. 1952. Intakes and retentions of nitrogen, calcium and phosphorus by 136 women between 30 and 85 years of age. Fed. Proc. 11:775–783.

Owen, G. M., P. J. Garry, K. M. Kram, C. E. Nelsen, and J. M. Montalvo. 1969. Nutritional status of Mississippi preschool children—A pilot study. Am. J. Nutr. 22:1444–1458.

Patwardhan, V. N. 1964. Protein–calorie deficiency disease: Public health aspects. Pages 310–323 in Proc. 6th Int. Congr. Nutr. Edinb., 1963. E. & S. Livingstone Ltd., Edinburgh.

Rao, B. S. N., K. V. Rao, and A. N. Naidu. 1969. Calorie–protein adequacy of the dietaries of pre-school children of India. J. Nutr. Diet. India. 6:238–244.

Swanson, P., E. Willis, L. M. Burrill, A. Biester, E. Jebe, M. A. Ohlson, and J. Smith. 1959. Food intakes of 2189 adult women in 5 North Central States. Iowa Agric. Exp. Stn. Res. Bull. 468 (North Cent. Reg. Publ. 83).

USDA (U.S. Department of Agriculture). 1964. The world food budget, 1970. Foreign Agric. Econ. Rep. No. 19. Economic Research Service, Foreign Regional Analysis Division. U.S. Government Printing Office, Washington, D.C. 105 pp.

USDA (U.S. Department of Agriculture). 1968. Food consumption, prices, expenditures. Agric. Econ. Rep. No. 138. Economic Research Service. 193 pp.

USDA (U.S. Department of Agriculture). 1969. Dietary levels of households in the United States, spring 1965. Rep. No. 6. Agricultural Research Service. U.S. Government Printing Office, Washington, D.C. 117 pp.

USDA (U.S. Department of Agriculture). 1970. National food situation. NFS–134. Economic Research Service. 32 pp.

USDA (U.S. Department of Agriculture). 1971. 1970 Suppl. Agric. Econ. Rep. No. 138. Economic Research Service. 91 pp.

USDA (U.S. Department of Agriculture). 1972. Food and nutrient intake of individuals in the United States, spring 1965. Rep. No. 11. Agricultural Research Service. U.S. Government Printing Office, Washington, D.C. 291 pp.

USDHEW (U.S. Department of Health, Education, and Welfare). 1972. Ten-state nutrition survey, 1968–1970. Publ. (HSM) 72-8130-8134. Washington, D.C.

U.S. Interdepartmental Committee on Nutrition for National Defense. 1963. Union of Burma, Nutrition Survey, Oct.–Dec. 1961. U.S. Government Printing Office, Washington, D.C. 287 pp.

Waterlow, J. C. 1968. Observations on the mechanism of adaptation to low protein intakes. Lancet 2:1091–1097.

A. E. HARPER and D. M. HEGSTED

Improvement of Protein Nutriture

One encounters two distinct problems in considering improvement of protein nutriture: First, improvement of the nutritional value of proteins in foodstuffs and diets; second, improvement of the nutritional status of individuals, communities, and populations. The former is relatively simple and subject to straightforward solutions; the latter is highly complex and does not have, nor is likely to have, a simple, single solution. All too frequently, the solution to the first of these problems is mistakenly assumed to be the solution to the second.

IMPROVING THE QUALITY OF FOOD PROTEINS

More than a half-century ago, Osborne and Mendel (1914) demonstrated that the addition of lysine to diets based on wheat proteins improved their nutritional value for the rat. Since then repeated animal experiments have demonstrated that the protein value of diets consisting mainly of cereal grains can be improved by supplements of amino acids, especially of lysine, and by supplements of lysine-rich proteins; that the protein quality of diets in which legumes (bean, pulse) are the main protein source can be improved by supplements of sulfur-containing amino acids or proteins rich in them; and that proteins with complementary amino acid patterns can be combined to produce a protein mixture of higher nutritional value than any of the individual component proteins (Hart, 1952; Flodin, 1953; Waddell, 1958; F N B , 1959; Rosenberg, 1959).

Also, experiments involving man have provided evidence that the nutritional value, or efficiency of utilization, of several plant proteins can be improved by appropriate supplements of amino acids or by combining them with proteins that have complementary amino acid patterns. The protein quality of diets consisting largely of wheat and wheat products for both adults (Hoffman and McNeil, 1949; Cremer *et al.*, 1951; Bricker *et al.*, 1945) and infants (Bressani *et al.*, 1960, 1963a; Barness *et al.*, 1961; Daniel *et al.*, 1968; Graham *et al.*, 1969) can be improved by supplements of lysine; that of diets consisting largely of maize, by supplements of lysine and tryptophan (Gomez *et al.*, 1957; Bressani *et al.*, 1958, 1963b; Truswell and Brock, 1961); that of diets consisting largely of sorghum and millet (Daniel *et al.*, 1965, 1966), or rice, by supplements of lysine and threonine (Parthasarathy *et al.*, 1964); and that of diets in which the protein is largely from soybeans and mixtures of beans or peanut and other foods, by supplements of methionine (Goyco, 1959; Nicol and Phillips, 1961; Graham, 1971b). These studies, which were carried out under the controlled conditions provided by a metabolic unit, have demonstrated that nitrogen retention of experimental subjects fed relatively low levels of these proteins is improved when the diets are supplemented with the limiting amino acids.

Reliable methods are available for determining the amino acid composition and overall digestibility of proteins, but methods for determining the availability of individual amino acids in foods are less well developed (see D. M. Hegsted, pp. 64–88 in this volume). Nevertheless, once the amino acid composition and digestibility of a low-quality protein have been established, the quality of products containing such proteins can then be improved by fortifying them with the amino acids in which they are low or with a quantity of protein that is rich in those amino acids, provided the availability of the amino acids pose no special problems, such as is sometimes encountered in severely heat-treated products. Specific technical problems may arise in fortifying complex foodstuffs with amino acids or with quantities of proteins having complementary amino acid patterns; but, as the two procedures are nutritionally equally efficacious, selection between them will hinge upon economic and technological considerations. These procedures have been used for many years to produce feedstuffs that are used efficiently by animals; the same principles apply in developing protein concentrates for human consumption (Scrimshaw *et al.*, 1959a).

These accomplishments do show that it is possible to increase the efficiency of utilization of food proteins and to develop products of high-protein quality from starting materials of low quality, but they bear only marginally on the problem of improving the nutritional status of man.

THE PROBLEM OF PROTEIN INADEQUACY

Undernutrition and malnutrition are major health problems in economically less favored countries; even in rich countries, segments of the population may be undernourished as a result of social instability, low income, limited nutritional knowledge, or inadequate health care. There is little evidence that protein malnutrition occurs in the United States, except as the result of ignorance, neglect, or consumption of inadequate amounts of food. Even in less well-to-do countries, protein–energy malnutrition, other than that due to an inadequate supply of food, is primarily a disease of young children. Hollingsworth and Greaves (1970) quote W. M. Bayliss, who has said of the Western European-type diet: "Take care of the calories and the protein will take care of itself." They go on to point out that "this condition is met for adults almost everywhere in the world. In other words, almost every mixture of foods liable to be eaten by adults, when eaten in sufficient quantity to meet their energy needs, is likely to meet protein needs." This viewpoint has been reiterated by the FAO/WHO/UNICEF Protein Advisory Group (1970). Calculations of the amino acid intakes of adults with low incomes in the United States (see H. H. Williams *et al.*, pp. 23–63, and E. F. Phipard, 167–183, in this volume) indicate that even those with low-protein intakes consume the recommended amounts of essential amino acids and that any inadequacy is likely to be associated with low-food intake. Calculations of the amounts of amino acids provided by plant products in quantities that supply the necessary amount of energy also lead to the conclusion that, for those above the age of 12 to have an inadequate intake of protein, they would have to subsist on such low-protein products as plantain, cassava, sweet potatoes, or yams (Autret *et al.*, 1968; Sukhatme, 1970a,b; Nicol, 1971) or be consuming a diet in which the amino acids were particularly unavailable (see D. M. Hegsted, pp. 64–88 in this volume).

Data on carbohydrate and fat consumption provide insight as to the adequacy of diets with regard to protein (Périssé, 1968; Périssé *et al.*, 1969). Animal products tend to be high in fat and plant products high in carbohydrate; as national income decreases and as individual incomes decrease within a given society, carbohydrate consumption tends to rise, owing to the greater dependence on the less-expensive plant products. The proportion of energy from protein does not fall appreciably, but the total amount of food that must be consumed to meet energy and protein requirements increases. Although the adult can consume enough of such foods to meet his protein requirements, the young child, with higher requirements per unit of body weight and a limited capacity for consumption, may not be able to eat enough of some of the bulky, high-carbohydrate products to obtain adequate amounts of either protein or energy. The problem is most acute in communities where the major staples are cassava and plantain (Nicol, 1971). Thus the young growing child is most at risk during periods of famine or food shortage, or when a low standard of living or a low family income forces him to depend upon inexpensive, high carbohydrate, low-protein foods for survival (Scrimshaw et al., 1969).

ETIOLOGY OF PROTEIN-ENERGY MALNUTRITION

If childhood malnutrition were due solely to dietary protein deficiency, providing protein supplements or improving the quality of the protein in dietary staples would solve the problem. However, the etiology of protein malnutrition is such that it is doubtful whether simple, uncomplicated protein deficiency occurs naturally to any significant extent; and there is no evidence for the occurrence of specific deficiencies of individual amino acids (see G. G. Graham, pp. 109–137 in this volume). Whitehead (1969) points out that Cicely Williams has been insisting since 1950 "that kwashiorkor is not just protein deficiency but many other things as well" and that a step toward recognizing this is the substitution, now widely accepted, of the term "protein–calorie" or "protein–energy" malnutrition for "protein malnutrition." He warns, nevertheless, that this, too, is an oversimplification.

Although protein deficiency has received major attention as the cause of worldwide malnutrition, the complexity of the disease has been emphasized frequently. Autret and Béhar (1954) based their preference for the term *sindrome policarencial infantil* on observations that "when the syndrome occurs among poor people, the protein deficiency is almost always accompanied by a deficiency of calories and various important nutrients." Multiple nutritional deficiencies have been recognized regularly as complicating factors in "protein malnutrition" (Hansen and Brock, 1954; Béhar et al., 1958; Whitehead, 1969). Scrimshaw and associates (1957) concluded that "protein malnutrition . . . is usually associated with a recent or long-standing deficiency of calories." This is reemphasized by Whitehead (1969), even though he points out that, where the habitual food is plantain, "a fairly pure form of protein malnutrition" may occur. Both Gopalan (1968, 1969) and Sukhatme (1970a,b) report, on the basis of dietary studies conducted in India, that "protein malnutrition" is almost invariably associated with a deficit of energy. In view of the well-recognized reduction in nitrogen retention that occurs when energy intake is inadequate (Calloway and Spector, 1954), it is not difficult to understand why "protein malnutrition" can occur as the result of a food deficit even when the food supply is not deficient in protein. Thus, from the viewpoint of diet alone, "protein malnutrition" is "many . . . things," as emphasized by Cicely Williams.

Beyond this, discussions of the etiology of "protein malnutrition" emphasize the importance of social customs and patterns. The use and distribution of foods are influenced by traditional social observances, superstititions, and taboos, frequently to the detriment of the young child. Even if sufficient high-quality food is available for the family, children may not receive as adequate a diet as do adults (Collis and Janes, 1968) (see also E. F. Phipard, pp. 167–183 in this volume). Owing to ignorance or custom, the child's food may be greatly diluted after weaning and during illnesses so that its energy and nutrient content are low (Whitehead, 1969). Furthermore, food intake commonly falls during episodes of diarrhea and the already-depleted child suffers further nutritional insult at just the time when needs are high owing to incomplete absorption of nutrients and wastage of tissues. The importance of infections and parasitic infestations as factors contributing to the development of protein malnutrition has been strongly emphasized (Scrimshaw *et al.,* 1959b; Bengoa, 1969). Infections can increase both energy requirements and nitrogen losses. Thus protein–energy malnutrition can occur even when the food supply is not conspicuously inadequate, either because of poor utilization in the home of what food is available, or because of poor utilization in the body as a result of nonnutritional diseases.

Hegsted (1970) has emphasized the difficulty of demonstrating clear-cut deficiencies of either protein or energy because of the close metabolic and nutritional relationships between the two and because a deficiency of one cannot be alleviated without affecting the other. He points out that treatment of protein–energy malnutrition involves the provision of a complete diet and that, therefore, one cannot establish which of the dietary components is of prime importance. This is obviously unlike the situation with deficiencies of many trace nutrients, in which specific responses to the administration of minute amounts of the appropriate nutrient can be demonstrated.

EXPERIENCE WITH FEEDING TRIALS

In an early review of kwashiorkor, Woods (1951) emphasized that an outstanding feature of that disease was its extraordinary resistance to therapy with such individual nutrients or semipurified substances as vitamins, lipotropic substances, liver extract, or protein concentrates. Limited evidence from general feeding trials with amino acid and protein supplements does not offer much promise that the relatively simple procedure of fortifying basic foodstuffs with amino acids, or even providing protein supplements, will alleviate protein–energy malnutrition. Widdowson and McCance (1954) fed some 300 undernourished children, aged 4–15 years, diets in which about 75 percent of the total energy was provided by bread and most of the remainder by vegetables. Their subjects' per capita intake of animal protein was about 8 g per day. The children grew throughout the 1-year period of the test at a rate considerably greater than well-nourished English or American children would have done and were judged to be well nourished. In a further test, the children of one group each received a pint of milk daily, and the children of a second group received an equicaloric amount of biscuits and orange juice. No differences between the two groups ascribable to diet variables were observed during the subsequent 6 months (McCance and Widdowson, 1955). Thus, with ample food, good sanitation, and appropriate health care, a diet composed largely of wheat products and vegetables was adequate for children aged 4 or older.

In a field trial in Haiti, malnourished school children fed a supplement of bread or one of bread fortified with lysine exhibited no clear-cut benefits from lysine fortification (King *et al.*, 1963). In another study in India, which utilized 2–5-yr-old children who received 54 percent of their calories and 85 percent of their protein from wheat, those subjects in a group receiving additional lysine grew, on the average, 0.6 cm more in 4 months than did those not receiving lysine (Pereira *et al.*, 1969). However, the two groups showed no difference in weight gain, nitrogen retention, serum protein concentration, or hemoglobin concentration; and it should be noted that the diet contained oil and sugar, so the protein content represented just 8 percent of the total energy. In a study of lysine fortification in Iran, no benefit from a lysine supplement was observed (Darby, 1970). In Peru, the effectiveness of a wheat–noodle supplement for children was compared to that of the same supplement fortified with fish protein concentrate (Graham *et al.*, 1963). Although the provision of additional wheat–noodle supplement enhanced nutritional status, the supplement containing fish protein was no more effective. In a field trial in Guatemala (Scrimshaw *et al.*, 1969), the administration of a food supplement to the population of one village "produced a measurable, although not a striking, result in promoting physical growth and development, and a lesser disease incidence" over that of a village not receiving the food supplement. Thus the provision of food, unaccompanied by other measures, brought about only a limited benefit.

ANALYSIS OF THE PROBLEM OF PROTEIN MALNUTRITION

A striking feature of the literature on what is commonly called "protein malnutrition" is the dearth of evidence that this syndrome results from any specific amino acid deficiency or even from simple protein deficiency. The early reports characterized the disease as malignant malnutrition of childhood or *sindrome policarencial infantil,* without focussing on any single causative agent. As information accumulated, two clinical entities were identified—one, marasmus, was equated with primary energy deficiency or total food deficiency; and the other, kwashiorkor, with primary protein deficiency (see G. G. Graham, pp. 109–137 in this volume). Despite recognition of these two extreme conditions, studies in India (Gopalan, 1969) suggest that differences in diet composition are probably less important than are other factors in determining which develops. The occurrence of protein deficiency, kwashiorkor, thus appears to be the result, not of a unique diet, but rather of some acute insult superimposed on a barely adequate or inadequate diet. In any event, the majority of cases appear to fall between the two conditions and are considered to represent varying degrees of deficiency of both energy and protein. Investigation of the causes of the syndrome and experience in treating it over the years have provided a large body of knowledge that reemphasizes its complexity and the futility of expecting that a simple, expedient solution for the problem can be found (Bengoa, 1965; Graham, 1965; Williams, 1965).

A recent survey of nutritional status in the United States has produced little evidence that protein deficiency is a general public health problem, despite the fact that a few cases of protein–energy malnutrition were identified (USDHEW, 1972). Food intake studies of children and families support this conclusion (see E. F. Phipard, pp. 167–183 in this volume). As noted above, when protein–energy malnutrition does occur, it appears to be the result of ignorance, neglect, or an inadequate intake of food associated with poverty. These, together with inadequate sanitation, widespread infections, and a general lack of health services and nutritional knowledge, are important causes of the worldwide problem of protein–energy malnutrition. Although there is no question but that the disease has a nutritional component and that it occurs primarily in areas with a low-protein, high-carbohydrate diet, it is not simply a nutritional problem, but rather a social and economic problem as well. It thus becomes important to assess critically the effectiveness and feasibility of apparently simple measures proposed for its alleviation.

Among the proposals that have received considerable attention in the United States and by the FAO/WHO/UNICEF Protein Advisory Group are amino acid fortification of basic foodstuffs, especially cereal grains, and the development of protein concentrates that can be used in small quantity to improve the household supply of protein. A more general approach to the world food problem has been developing slowly over the past 25 yr, i.e., the improvement of agricultural production through genetic improvement of staple food crops and the application of modern agricultural methods—the so-called "green revolution." This last is recognized as a long-term solution and, as such, requires a long-term commitment by governments and by well-trained individuals. The first two proposals are put forward as interim solutions to the immediate problem, while the longer-term solution is being developed.

To put into perspective the proposal that widespread fortification of cereal grains with amino acids be initiated, it must be acknowledged at the outset that the protein quality of diets containing proteins with unbalanced patterns of amino acids can be improved by supplements of the limiting amino acids. It must also be recognized that, with the exception of a few diets based largely on root crops or plantain, unsupplemented low-protein diets will meet the protein requirements of adults who consume sufficient food to meet their energy needs. Blix (1965), in a study of human diets, found that the relationship between the intake of energy and the intake of protein tended to remain constant and that low-protein intakes were almost invariably associated with an energy deficit. Even a diet in which 90–95 percent of the nitrogen was supplied by white flour was sufficient to maintain 19–27-year-old men who consumed an adequate amount of energy in nitrogen equilibrium (Bolourchi et al., 1968). The studies with children who were fed a diet in which 75 percent of the protein was from bread demonstrated that children aged 4–15 years can grow well on diets with relatively unbalanced amino acid patterns (Widdowson and McCance, 1954). Begum et al. (1970) have reported that children aged 2–5 fed a diet composed entirely of plant products (100 kcal/kg), with 83 percent of the dry matter derived from rice, wheat, and nonprotein energy sources, grew at the same rate as did North American children in the 50th percentile. Although protein–energy malnutrition is a problem of the young child, it is most severe in the young child who receives insufficient food and is subject to recurring infections.

Despite the fact that general amino acid fortification of staple foods could, at best, be expected to benefit only a small part of the population, primarily very young children, the procedure could be justified if its efficacy were assured. However, it has not yet been established that measurable benefits can be obtained in a population through fortification of staple foods with amino acids. Hegsted (1968) assessed the effectiveness of amino acid fortification as a means of improving the nutritional quality of wheat flour and concluded that the improvement from lysine and threonine supplements is much less than is often predicted. Furthermore, Miller and Donoso (1963) concluded, from biological assays of food mixtures, that in many poor countries, despite a high intake of cereal grains, the sulfur-containing amino acids are likely to be the limiting amino acids. The FAO/WHO/UNICEF Protein Advisory Group (1970) states as its first consideration for recommending amino acid fortification that the limiting amino acid in the diet of the target group should be established and that energy intake should be adequate. Presumably, other dietary deficiencies should be corrected as well. Amino acid fortification, while it may improve the protein quality of a staple foodstuff, contributes only trivially to the total energy intake and does not alleviate any other existing dietary deficiencies. Also, questions arise as to whether it will improve the protein quality of the diet as a whole; and, even if it does, whether this improvement will be reflected in improved nutritional status if energy intake remains low and if infections and illnesses remain as complications in the target population.

Quite apart from nutritional considerations, economic decisions concerning how much of a limited budget should be invested in a specific program must be made. Much has been said about the utility of cost-benefit analysis in deciding among various procedures for improving protein nutrition. What is overlooked in such considerations is that cost-benefit analysis is meaningless if a benefit per se cannot be established and quantified. Because evidence of direct benefit deriving from amino acid fortification is lacking, calculations are based on anticipated theoretical improvement in protein quality (Hegsted, 1968). A compilation of cereal grain consumption of some 17 Asian, African, and Latin American countries shows that the percent of total per capita protein consumption from grains ranges between 40–71 percent (Kracht, 1969). How closely the theoretical nutritional value of the total dietary protein approaches the actual value will, therefore, depend upon the amino acid composition of the other 29–60 percent of protein in the diet.

Thus, the basis for calculating benefit is tenuous at best. Further, if the cost of the amino acid alone is used in calculating the increase in price to the consumer, the increase for wheat fortified with 0.2 percent of lysine is about 2.5 percent (Hegsted, 1968). However, a more detailed analysis of the total cost indicates that the basic price of flour fortified in this way would be increased by 6–8 percent (Kracht, 1969). Inasmuch as cereal grains often represent from 40 to 70 percent of a national food supply, this increase would represent a substantial sum of money, especially if foreign exchange had to be used to purchase it—as indeed it would until such time as local plants were developed to produce amino acids. For a procedure of unproven merit, this appears to be a high-risk venture.

The development of protein mixtures fortified with vitamins and minerals has progressed in many areas. Such products are commercially available in several developing countries. They are designed so that relatively small amounts included in the food during cooking will improve the total nutritional quality of the diet. This procedure has more nutritional merit as a means of improving individual diets than does amino acid fortification of staple foods in that it provides additional quantities of several essential nutrients as well as protein and some energy.

The efficacy of fortified protein mixtures, however, will be limited if energy intake is inadequate. Unless they are consumed during periods of illness, when food intake is commonly reduced, the benefits they can provide may be lost at the most critical time. Perhaps most important, if they are not available at a cost that permits them to compete with existing staple foods that satisfy the need for energy, such staples are likely to take precedence in the food budget of poor families. To assure their distribution to those who might most benefit from them will probably require government subsidization. Thus, although the nutritional basis for the development of protein mixtures fortified with vitamins and minerals is sound, the potential of such programs for alleviating malnutrition is limited.

Analyses of the problem of improving protein nutrition (Autret, 1970; Hollingsworth and Greaves, 1970) emphasize that the most important measure to increase protein supply is to increase the production of cereal grains and that improvement of the quality and the quantity of proteins of the grains will further improve the supply of protein. Because in many regions cereals provide the most energy for the least cost, they often constitute the major food supply of poor people and developing countries. Thus increased production provides greater supplies of both energy and protein, as well as of many other nutrients for the population as a whole. If the extra food is readily available, the major problem lies with the young child, whose nutrient requirements are higher than those of the adult. However, because the child's energy needs are also higher, requirements for most nutrients per unit of energy are not as much greater as it may appear when requirements are expressed per unit of body weight. Nevertheless, this vulnerable group should be given special attention. The British food program during and after World War II (Hollingsworth and Greaves, 1970), which focussed especially on the needs of young children and pregnant and nursing women, is of special interest in this regard. The program centered on providing milk as the food supplement for those groups; and, although in most developing countries milk in the required amounts and of adequate quality is not available, other types of high-quality foods especially designed as weaning foods for children can be a valuable part of an effective national food program (Graham, 1971a; Nicol, 1971). It is in the development of special high-quality foods for such programs that amino acid and protein supplements can be most useful. Again, initiation of such a program is not just a nutritional problem; it can be accomplished only by commitment, both financial and educational, on the part of the appropriate government agencies, to a carefully designed national food program. Piecemeal approaches offer little to commend.

The same level of commitment is essential for improving the total supply of protein and food. Borlaug (1968) has reviewed the development of the wheat program in Mexico. Only through such long-term efforts can any lasting solution to the food and protein problems be achieved. The culmination of the program in Mexico required 20 yr of effort through agronomic research, the development of an extension service program, the training of local scientists, organization of methods of financing farmers to enable them to acquire new seed and fertilizer, and much more. Nevertheless, it successfully converted Mexico from a wheat importing nation to one that is self-sufficient, increased the income of many farm units, and resulted in a saving of foreign exchange for national investment. Tuck (1970) has emphasized the prime importance of concentrating on food production, rather than on other industries, as the first step in economic development. Yet, even in Mexico it is not yet clear how much improvement in the nutritional status of the population has been achieved. Improvement of nutritional status depends upon measures other than increased productivity alone; in Asia, for example, the higher yields from the "Green Revolution" apparently have not benefitted those most in need (Brown and Rastyannikov, 1971).

Whenever the causes of protein–energy malnutrition include inadequate energy supply, inadequate supply of total food, inappropriate distribution of the available food, ignorance of basic nutrition knowledge, unsanitary living conditions, insufficient numbers of appropriately educated personnel, lack of medical care, low income, and a slow rate of economic development, no simple solutions will likely be found. To advocate simple expedients as solutions for complex problems not only does little to alleviate the basic difficulty, but also obscures the real issues; delays the development of the adequate, long-term solutions that are essential; and tends to divert funds and effort into directions that are likely to delay rather than promote permanent solutions.

Critical analysis leads to the conclusion that protein–energy malnutrition intimately relates to poverty and that no general formula exists for solving the basic social and economic problems in which worldwide malnutrition has its roots. Unique conditions are encountered in different countries and even in different localities within a country. To ensure that a program will benefit a population that is at risk of malnutrition requires systematic analysis of each situation and critical assessment of each proposed solution. It is particularly important to identify accurately the basic causes of malnutrition, to develop a strategy that focuses specifically on the needs of the target population and to evaluate the effectiveness of the program continuously so it can be modified as its weaknesses and strengths are recognized. Without such an assessment, despite the highest of motives, programs that are inappropriate or ineffective will undoubtedly be undertaken. For example, lysine fortification can improve the nutritional value of wheat proteins but is of no value if the overall diet is deficient in neither lysine nor protein; it is of limited value even when lysine supply is inadequate, if it raises the cost of a staple produce without at the same time increasing the purchasing power of those who need the supplement; introduction of nutritionally improved cereals will be without benefit and can even be detrimental if the yield is so low that total energy production falls. These relationships should be obvious even to the layman; however, many superficially simple proposals for changes in agricultural practices or in the food supply may have subtle and far-reaching ramifications that are not readily recognized without careful analysis of the local situation.

Increased food production and a national food program are two critical steps in alleviating malnutrition; they are but two among many. For nutrition programs to be effective, they must be coupled with education and public health programs. Limited reports of success in reducing malnutrition through having trained dietitians provide mothers with nutritional information (Collis and Janes, 1968) indicate the potential for educational programs. The success of rehabilitation (mothercraft) centers (King, 1972) supports this view. The importance of disease in precipitating bouts of malnutrition is well documented and emphasizes the need for public health measures directed toward improved sanitation, water supplies, and community health services. Full institution of these programs and the ultimate solution of the problem of worldwide malnutrition, if it is to be achieved, undoubtedly depend in the final analysis upon economic development. Initiation of health and agricultural programs directed toward alleviating malnutrition can proceed concurrently with, and can contribute to, economic development. Nevertheless, even the most effective nutrition programs will meet with only limited success unless economic development proceeds steadily and is accompanied by an increase in the purchasing power of those with bare subsistence incomes.

REFERENCES

Autret, M., and M. Béhar. 1954. Sindrome policarencial infantil (kwashiorkor) and its prevention in Central America. FAO Nutr. Stud. No. 13. FAO, Rome. 7 pp.

Autret, M. 1970. World protein supplies and needs. Pages 3–19 in R. A. Lawrie, ed. Proteins as human food. Butterworths, London.

Autret, M., J. Pérrisé, F. Sizaret, and M. Cresta. 1968. Protein value of different types of diet in the world. FAO Nutr. Newsl. 6(4):1–29.

Barness, L. A., R. Kay, and A. Valyaseve. 1961. Lysine and potassium supplementation of wheat protein. Am. J. Clin. Nutr. 9:331–344.

Begum, A., A. N. Radhakrishnan, and S. M. Perreira. 1970. Effect of amino acid composition of cereal-based diets on growth of pre-school children. Am. J. Clin. Nutr. 23:1175–1183.

Béhar, M., F. Viteri, R. Bressani, G. Arroyave, R. L. Squibb, and N. S. Scrimshaw. 1958. Principles of treatment and prevention of severe protein malnutrition in children (kwashiorkor). Annu. N.Y. Acad. Sci. 69:954–968.

Bengoa, J. M. 1965. The prevention of malnutrition in young children. Pages 36–43 in Proc. West. Hemisphere Nutr. Congr. AMA, Chicago.

Bengoa, J. M. 1969. Outline of WHO and FAO participation in research on protein-calorie malnutrition. Pages 10–18 in A. Van Murall, ed. Protein–calorie malnutrition. Springer–Verlag, Berlin.

Blix, G. 1965. A study on the relation between total calories and single nutrients in Swedish food. Acta Soc. Med. Upsala 70:117–129.

Bolourchi, S., C. M. Friedemann, and O. Mickelsen. 1968. Wheat flour as a source of protein for adult human subjects. Am. J. Clin. Nutr. 21:827–835.

Borlaug, N. E. 1968. Wheat breeding and its impact on world food supply. Pages 1–36 in Proc. 3rd Int. Wheat Genet. Symp. Aust. Acad. Sci., Canberra.

Bressani, R., N. S. Scrimshaw, M. Béhar, and F. Viteri. 1958. Supplementation of cereal proteins with amino acids. II. Effect of amino acid supplementation of corn-masa at intermediate levels of protein intake on nitrogen retention of young children. J. Nutr. 66:501–513.

Bressani, R., D. Wilson, M. Béhar, M. Chung, and N. S. Scrimshaw. 1960. Supplementation of cereal proteins with amino acids. III. Effect of amino acid supplementation of wheat flour as measured by nitrogen retention in young children. J. Nutr. 70:176–186.

Bressani, R., D. Wilson, M. Chung, M. Béhar, and N. S. Scrimshaw. 1963a. Supplementation of cereal proteins with amino acids. IV. Lysine supplementation of wheat flour fed to young children at different levels of protein intake in the presence and absence of other amino acids. J. Nutr. 79:333–339.

Bressani, R., D. Wilson, M. Chung, M. Béhar, and N. S. Scrimshaw. 1963b. Supplementation of cereal proteins with amino acids. V. Effect of supplementing lime-treated corn with different levels of lysine, tryptophan, and isoleucine on the nitrogen retention of young children. J. Nutr. 80:80–84.

Bricker, M., H. H. Mitchell, and G. M. Kinsman. 1945. The protein requirement of adult human subjects in terms of the protein contained in individual foods and food combinations. J. Nutr. 30:269–283.

Brown, L., and V. G. Rastyannikov. 1971. The green revolution, with extracts from *The Green Revolution, Rural Employment and the Urban Crisis* by L. Brown and *Socio-economic Aspects of the Green Revolution* by V. G. Rastyannikov. Pages 127–137 *in* B. Ward, J. D. Runnalls, and L. D'Anjou, eds. The widening gap; developments in the 1970's. Columbia University Press, New York.

Calloway, D. H., and H. Spector. 1954. Nitrogen balance as related to calorie and protein intake in active young men. Am. J. Clin. Nutr. 2:405–412.

Collis, W. R. F., and M. Janes. 1968. Multifactorial causation of malnutrition and retarded growth and development. Pages 55–71 *in* N. S. Scrimshaw and J. E. Gordon, eds. Malnutrition, learning, and behavior. The MIT Press, Cambridge, Mass.

Cremer, H. D., K. Lang, I. Hubbe, and U. Kulik. 1951. Versuche zur Aufbesserung der biologischen Wertigheit von Weizeneiweiss durch Lysin oder Hefe und ein Vergleich mit Hafereiweiss. Biochem. Z. 322:58–67.

Daniel, V. A., R. Leela, T. R. Doraiswamy, D. Rajalakshmi, S. Venkat Rao, M. Swaminathan, and H. A. B. Parpia. 1965. The effect of supplementing a poor Indian Ragi diet with L-lysine and DL-threonine on the digestibility coefficient biological value and net utilization of the protein and on nitrogen retention in children. J. Nutr. Diet. 2:138–143.

Daniel, V. A., R. Leela, T. R. Doraiswamy, D. Rajalakshmi, S. Venkat Rao, M. Swaminathan, and H. A. B. Parpia. 1966. The effect of supplementing a poor Kaffir corn (*Sorghum vulgare*) diet with L-lysine and DL-threonine on the digestibility coefficient, biological value and net utilization of protein and retention of nitrogen in children. J. Nutr. Diet. 3:10–14.

Daniel, V. A., T. R. Doraiswamy, S. Venkat Rao, M. Swaminathan, and H. A. B. Parpia. 1968. The effect of supplementing a poor wheat diet with L-lysine and DL-threonine on the digestibility coefficient, biological value, and net utilization of proteins and nitrogen retention in children. J. Nutr. Diet. 5:134–141.

Darby, W. J. 1970. Nowhere is too far. Nutr. Today 5:32–33.

FAO/WHO/UNICEF PAG recommendation on amino acid fortification of foods. 1970. PAG Statement No. 9.

Flodin, N. W. 1953. Amino acids and proteins; their place in human nutrition problems. J. Agric. Food Chem. 1:222–235.

FNB (Food and Nutrition Board, National Research Council). 1959. Evaluation of protein nutrition. Publ. 711. National Academy of Sciences, Washington, D.C. 61 pp.

Gómez, F., R. Ramos-Galván, J. Cravioto, S. Frenk, C. de la Peña, M. E. Moreno, and M. E. Villa. 1957. Protein metabolism in chronic severe malnutrition (kwashiorkor). 2. Influences of aminoacid supplements on the absorption and retention of nitrogen from a maize and beans diet. Acta Paediat. 46:286-293.

Gopalan, C. 1968. Kwashiorkor and marasmus. Evolution and distinguishing features. Pages 49-69 in R. A. McCance and E. M. Widdowson, eds. Calorie deficiencies and protein deficiencies. Little, Brown and Co., Boston.

Gopalan, C. 1969. Observations on some epidemiological factors and biochemical features of protein–calorie malnutrition. Pages 77-85 in A. von Muralt, ed. Protein calorie malnutrition. Springer–Verlag, Berlin.

Goyco, J. A. 1959. Nitrogen balance of young adults consuming a deficient diet supplemented with Torula yeast and other nitrogenous products. J. Nutr. 69: 49-57.

Graham, G. G. 1965. Methods of combating malnutrition. Pages 60-63 in Proc. West. Hemisphere Nutr. Congr. AMA, Chicago.

Graham, G. G. 1971a. Feeding trials in children. Pages 358-364 in G. F. Stewart and C. L. Willey, eds. Proc. 3rd Int. Congr. Food Sci. Tech., 1970. Institute of Food Technologists, Chicago.

Graham, G. G. 1971b. Methionine or lysine fortification of dietary protein for infants and small children, p. 222-236. Rep. Int. Natl. Conf. Amino Acid Fortification Protein Foods, MIT, Cambridge, Mass., 16-18 Sept. 1969. Pages 222-236 in N. S. Scrimshaw and A. M. Altschul, eds. Amino acid fortification of foods. The MIT Press, Cambridge, Mass.

Graham, G. G., A. Cordano, and J. M. Baertl. 1963. Studies in infantile malnutrition. II. Effect of protein and calories intake on weight gain. J. Nutr. 81:249-254.

Graham, G. G., R. P. Placks, G. Acevedo, E. Morales, and A. Cordano. 1969. Lysine enrichment of wheat flour: Evaluation in infants. Am. J. Clin. Nutr. 22:1459-1468.

Hansen, D. L., and J. F. Brock. 1954. Potassium deficiency in the pathogenesis of nutritional edema in infants. Lancet 2:477.

Hart, E. B. 1952. Abuse of data on the biologic value of proteins. Nutr. Rev. 10: 129-130.

Hegsted, D. M. 1968. Amino acid fortification and the protein problem. Am. J. Clin. Nutr. 21:688-692.

Hegsted, D. M. 1970. Malnutrition in developing countries. Seminar on Food Problems in Asia and the Pacific. Honolulu, Hawaii.

Hoffman, W. S., and G. C. McNeil. 1949. Enhancement of the nutritive value of wheat gluten by supplementation with lysine, as determined from nitrogen balance indices in human subjects. J. Nutr. 38:331-343.

Hollingsworth, D. F., and J. P. Greaves. 1970. Nutrition policy with regard to protein. Pages 32-45 in R. A. Lawrie, ed. Proteins as human food. Proc. 16th Easter School Agric. Sci., Univ. Nottingham, 1969. AVI Publishing Co., Westport, Conn.

King, K. W. 1972. Nutrition education of the poor can be effective. Indian J. Nutr. Diet. 9:351-356.

King, K. W., W. H. Sebrell, E. L. Severinghaus, and W. O. Storvick. 1963. Lysine fortification of wheat bread fed to Haitian school children. Am. J. Clin. Nutr. 12:36-48.

Kracht, V. 1969. Economic aspects of the supplementation of cereals with lysine. FAO/WHO/UNICEF/PAG Rep.

McCance, R. A., and E. M. Widdowson. 1955. Old thoughts and new work on breads white and brown. Lancet 2:205–210.

Miller, D. S., and G. Donoso. 1963. Relationship between the sulphur/nitrogen ratio and the protein value of diets. J. Sci. Food Agric. 14:345–349.

Nicol, B. M. 1971. Protein and calorie concentration. Nutr. Rev. 29:83–88.

Nicol, B. M., and P. G. Phillips. 1961. Reference groundnut flour (GNF) and reference dried skimmed milk (DSM) as supplements to the diets of Nigerian men and children. Pages 157–168 in Progress in meeting protein needs of infants and preschool children. Publ. 843. National Academy of Sciences, Washington, D.C.

Osborne, T. B., and L. B. Mendel. 1914. Amino acids in nutrition and growth. J. Biol. Chem. 17:325–349.

Parthasarathy, H. N., J. Kantha, V. A. Daniel, T. R. Doraiswamy, A. N. Sankaran, M. Narayana Rao, M. Swaminathan, A. Sreenevasan, and V. Subrahmanyan. 1964. The effect of supplementing rice diet with lysine, methionine, and threonine on the digestibility coefficient, biological value, and net protein utilization of the proteins and on the retention of nitrogen in children. Can. J. Biochem. 42:385–393.

Pereira, S. M., A. Begum, G. Jesudian, and R. Sundararaj. 1969. Lysine-supplemented wheat and growth of preschool children. Am. J. Clin. Nutr. 22:606–611.

Périssé, J. 1968. The nutritional approach in food policy planning. FAO Nutr. Newsl. 6(1):30–45.

Périssé, J., F. Sizaret, and P. Francois. 1969. The effect of income on the structure of the diet. FAO Nutr. Newsl. 7(3):1–9.

Rosenberg, H. R. 1959. Amino acid supplementation of foods and feeds. Pages 381–418 in A. A. Albanese, ed. Protein and amino acid nutrition. Academic Press, New York.

Scrimshaw, N. S., M. Béhar, G. Arroyave, C. Tejada, and F. Viteri. 1957. Kwashiorkor in children and its response to protein therapy. J. Am. Med. Assoc. 164: 101–118.

Scrimshaw, N. S., R. L. Squibb, R. Bressani, M. Béhar, F. Viteri, and G. Arroyave. 1959a. Vegetable protein mixtures for the feeding of infants and young children. Pages 28–46 in W. Cole, ed. Amino acid malnutrition. Rutgers University Press, New Brunswick, N.J.

Scrimshaw, N. S., C. E. Taylor, and J. E. Gordon. 1959b. Interactions of nutrition and infection. Am. J. Med. Sci. 237:367–403.

Scrimshaw, N. S., M. Béhar, M. A. Guzman, and J. E. Gordon. 1969. Nutrition and infection field study in Guatemalan villages, 1959–1964. IX. An evaluation of medical, social and public health benefits with suggestions for future field study. Arch. Environ. Health 18:51–62.

Sukhatme, P. V. 1970a. Size and nature of the protein gap. Nutr. Rev. 28:223–226.

Sukhatme, P. V. 1970b. Incidence of protein deficiency in relation to different diets in India. Br. J. Nutr. 24:477–487.

Truswell, A. S., and J. F. Brock. 1961. Effects of amino acid supplements on the nutritive value of maize protein for human adults. Am. J. Clin. Nutr. 9:715–728.

Tuck, R. H. 1970. Economics of protein production. Pages 20–31 in R. A. Lawrie, ed. Proteins as human food. Butterworths, London.

USDHEW (U.S. Department of Health, Education, and Welfare). 1972. Ten-state nutrition survey, 1968–1970. Publ. (HSM) 72-8130-8134.

Waddell, J. 1958. Supplementation of plant proteins with amino acids. Pages 307–352 *in* A. M. Altschul, ed. Processed plant protein foodstuffs. Academic Press, New York.

Whitehead, R. G. 1969. Factors which may affect the biochemical response to protein–calorie malnutrition. Pages 38–47 *in* A. Von Murall, ed. Protein–calorie malnutrition. Springer–Verlag, Berlin.

Widdowson, E. M., and R. A. McCance. 1954. Studies on the nutritive value of bread and on the effect of variations in the extraction rate of flour on the growth of undernourished children. Med. Res. Counc. (Br.) Spec. Rep. Ser. No. 287:1–137.

Williams, C. D. 1965. Factors in the ecology of malnutrition. Pages 20–24 *in* Proc. West. Hemisphere Nutr. Congr. AMA, Chicago.

Woods, R. 1951. Kwashiorkor—the syndrome of malignant malnutrition. Borden Rev. Nutr. Res. 12:35–46.

Contributors

Guillermo Arroyave, Chief, Division of Physiological Chemistry, Instituto de Nutricion de Centro America y Panama, Carretera Roosevelt Zona 11, Guatemala, C.A.

George G. Graham, Professor of International Health, Johns Hopkins Medical Institution, 615 North Wolfe Street, Baltimore, Md. 21205

Alfred E. Harper, Professor, Department of Nutritional Sciences and Biochemistry, University of Wisconsin, Madison, Wis. 53706

D. Mark Hegsted, Professor, Department of Nutrition, School of Public Health, Harvard University, 665 Huntington Avenue, Boston, Mass. 02115

L. E. Holt, Jr., Department of Pediatrics, New York University, 550 First Avenue, New York, N.Y. 10016

J. M. McLaughlan, Food and Drug Research Laboratories, Department of National Health and Welfare, Ottawa, Canada

Esther F. Phipard, 1831 Kirby Road, McLean, Va. 22101

Harold H. Williams, 1060 Highland Road, Ithaca, N.Y. 14850